YAMAHA

YAMAHA

Mick Walker

OSPREY
AUTOMOTIVE

First published in Great Britain in 1993
by Osprey, an imprint of Reed Consumer
Books Limited, Michelin House,
81 Fulham Road, London SW3 6RB and
Auckland, Melbourne, Singapore and Toronto

ISBN 1 85532 342 7

Editor Shaun Barrington
Page design Paul Kime/Ward Peacock
Partnership
Photography by Dennis Baldry, Terry Howe
and Mick Walker

Printed in Hong Kong

Front cover
Kenny Roberts, race number 3,
accelerating hard, chin on the tank to
minimise drag, aboard a factory 500 cc
YPVS. Roberts is wearing his distinctive
French Dainese racing leathers and an
Italian AGV helmet decorated with the
head of a North American Bald Eagle

Back cover
The watercooled 350LC was launched
in 1980 together with the 250LC; both
bikes were superbly versatile 2-stroke
sports roadsters

Half title page
Kenny Roberts in typical stylish action
on his works YZR500 four, summer
1983

Title page
New for 1993, the YZF750R brought
factory racer performance to the street.
As emphasised here, the immensely
strong and rigid Deltabox frame is the
key element in the bike's seemingly
telepathic handling qualities, while its
excellent aerodynamics and 749 cc
5-valve motor produce a maximum
speed potential of over 160 mph. A
generous 20-litre aluminium fuel tank
was required for race winning
endurance and gives the YZF750R a
highway cruising range of about
200 miles

Above
Ulsterman Mark Farmer with the FZR600 he rode at Silverstone, August 1992

Introduction

Yamaha is a word that conjures up a sporting image with any motorcycle enthusiast; though if its successes in road racing, motocross or trials are combined, both Honda and Suzuki could probably boast as many if not more victories!

The reason for this image lies in Yamaha's unwavering commitment to racing since the early 1960s, which has included both pukka works machinery and the offering of production 'over-the-counter' bikes for both club and national riders around the world.

Besides two-wheelers the company has also manufactured musical instruments (including some superb pianos), snowmobiles, industrial engines, marine engines, lawnmowers, unmanned helicopters, industrial robots and even swimming pools.

In many ways Yamaha can be compared with the Italian Ducati marque, because both are at their best offering specialised motorcycles with a distinctly sporting bias. Yamaha's biggest (and most costly) failures have come when trying to produce touring bikes — for example the TX750 parallel twin, the XS500 parallel twin, and the XZ550 and TR1 V-twins — not to mention the ill-fated XJ650 Turbo.

Conversely, Yamaha has created such classics as the RD250/350 LCs, TZR250, FZR 1000 and the superb YZF750 — all of these designs are uncompromised sportsters.

Right
VX535 Virago custom: V-twin reliability with soul-satisfying torque

Contents

Early Days

Born in Nagasaki in 1851, Torakusu Yamaha was 20 years old when he began a ten year apprenticeship under the guidance of a British clock maker. This was to provide an excellent basis for a working life in the field of precision engineering. Not content with this decade of learning, Yamaha then took up another apprenticeship at Osaka to make medical equipment. In 1883 he decided to move to the Hamamatsu area, and four years later – now self-employed as a general engineer – Torakusu was asked to repair an organ at the Hamamatsu Primary School.

It was this job that set Yamaha on a course that would eventually lead his company, Nippon Gakki, to become one of the world's foremost manufacturers of musical instruments. Before the turn of the century, the company was not only a major supplier on the home market, but had already begun an export drive, which included shipping some 80 organs to Britain in 1892.

Expansion continued unabated, but in 1916, Torakusu Yamaha died at the age of 64. By then, however, Nippon Gakki was firmly established. The next man to run the company was Chiymanu Amano, but a series of problems in the early 1920s, including strikes and the Kanto earthquake of 1923, created severe financial problems. These were only solved in 1926, when a new president, Kaichi Kawakami, was appointed.

Compared with Amano, whose dictatorial regime had largely created the unrest, Kawakami went out of his way to combine both the management and workers into an effective team – it worked and has since become standard practice in Japanese industrial relations.

After the fortunes of the company had been rebuilt, the rise to power of the military regime in Japan posed another problem for Kawakami during the late 1930s. By 1940, the army had taken control of the plant which, during the war years, manufactured aviation products, including fuel tanks and propellers. As for musical instrument production, this was allowed to continue to a very limited degree until 1944, when the authorities issued a total ban.

During early 1945 many of the Nippon Gakki production facilities were badly damaged by Allied bombing, and following the end of hostilities, the road back to corporate recovery was long and hard. However, by 1948, things were improving with the resumption of musical instrument manufacture. Another advantage was that owing to Kawakami's support and active help from his employees over the years, Nippon Gakki was unaffected by the series of strikes that brought several Japanese firms (including Suzuki) to their knees during this period.

Famous three tuning forks logo of the Yamaha Motor Co. It stems from the company's manufacture of musical instruments, which is still an important part of Yamaha today

During 1950, Kawakami, who had suffered from ill-health for some time, passed the presidency to his 38-year-old son, Genichi Kawakami. This was to prove the vital step towards the emergence of the Yamaha motorcycle marque. One of the younger Kawakami's first moves was to take a long, hard look at the company, and during his investigation he decreed that wartime machinery, locked away since the end of the conflict, should be utilised. The big question was for what purpose? In common with many of his industrial counterparts in Japan (and Germany and Italy), Kawakami decided to manufacture powered two-wheelers.

Unlike many, however, the Nippon Gakki president wanted, above all else, a quality product – one which would provide the same level of satisfaction as had already made the company's musical instruments and organs so respected.

The biggest problem was choosing the right type of machine. After much deliberation, the company decided to concentrate upon a motorcycle based around the highly successful German DKW RT125 design. This had first appeared in 1935, and in the post-war era was destined to be copied by not only by Nippon Gakki, but also BSA, Harley-Davidson and the Russian Voskhod factory, among others.

After an extremely comprehensive development programme and rigorous testing – including personal involvement in the latter by president Kawakami, who obviously liked to lead from the front – a new factory was built at Hamamatsu for the express purpose of producing the new machine. In addition, the enterprise was named in honour of the company's founder – hence Yamaha, rather than Nippon Gakki, was to adorn the new motorcycle's tank sides!

The reputation of this first machine, the YA1, was enhanced considerably by its domination of the Mount Fuji and Asama races in 1955. By the end of that year, some 200 YA1s were leaving the factory each month, and work had begun on a larger model. Again, this was based around a DKW – the RT175 – and appeared at the end of the year.

But the next machine was to provide Yamaha with a motorcycle of significant importance for the future, one that would enable it to compete with the established marques. Yamaha felt that the new bike had to be a 250. Originally, another DKW was considered as a base, but this was a single, so another German machine, a twin-cylinder Adler MB250, was imported for evaluation.

At the time, the Adler was considered by many to represent the ultimate in two-stroke development, so when the development team, headed by Zenzaburo Watase, informed president Kawakami that it preferred to follow its own design route, this was a bold move indeed. In retrospect, Kawakami must be applauded for approving such a bold move. The result was a trend setting machine (coded YD1), in which the only real Adler influence lay in the construction of the crankcase assembly – and, of course, that both were 54 × 54 mm piston-ported two-stroke parallel twins.

Another reason for the interest in the German design was its performance on the race track. As already related, Yamaha had won the two Japanese races held in 1955 – at Mount Fuji (the last of three) and Asama (the first to be run at this venue). Strangely, no races were held in 1956, but the second Asama event was scheduled to be staged in November 1957. With the debut of the new 250 twin in April that year, it was perhaps inevitable the Yamaha would choose to enter derivatives of the production bike.

However, Yamaha realised that certain features of the machine would need modification, most notably the pressed-steel frame. In its place was a much neater (and lighter) tubular duplex device. In addition, a very careful weight-pruning exercise was carried out on the power unit, together with additional engine tuning and a pair of expansion-chamber exhausts. The gear ratios inside the box were changed to make fuller use of the available 17 bhp, some 3 bhp more than the standard engine.

A new regulation called for the use of only one machine per capacity class, which had to be of the same specification for each team entry, so Yamaha took this opportunity to introduce a shorter-stroke variant of the new twin. This featured revised bore and stroke measurements of 56 × 50 mm, dimensions which were to be employed on Yamaha sports roadsters and racers in the future.

YAMAHA 100 TWIN
MODEL YL-1
with Yamaha Autolube
EST 1887

Discover the swinging world of YAMAHA

In the YL1 Yamaha have come up with a miniature masterpiece; a twin cylinder 100 c.c. engine which will hold its own with many machines of twice its engine size. It has twin matched carburettors, twin tuned silencers, separate twin cylinders and many engine parts which are made to aircraft engine limits. With its light frame and damped springing this new Yamaha is responsive and handles splendidly in dense traffic. It has a real "get away" when it reaches the open road.

Above

The tiny YL100 twin ran from 1966 through to 1971. Its 97 cc (38 × 43 mm) engine produced 9.5 bhp and 70 mph in standard trim. Yamaha also offered a comprehensive race kit which bumped maximum speed up to 95 mph (track use only), consisting of new heads, 5-port cylinders, windowed pistons, larger carbs and expansion chambers. With this fitted the power output nearly doubled to 18 bhp at 8500 rpm

Right
Long running YA series began in the
late 1950s and continued until the early
1970s. This is the YA6, like the rest of
the family it used rotary disc induction
and an inclined cylinder as part of its
specification

The machines which Yamaha entered for the 1957 Asama races were virtually works specials, built to win. President Kawakami saw the importance of victory and the need to defeat the company's biggest rival, Honda, on the circuit. This is exactly what it did, taking the first three places in the 250 cc race, and first and second in the 125 cc event.

As with their earlier success in the 1955 event, Yamaha found customers were influenced in their choice of a new machine by racing victories and, once again, the company experienced a surge in orders for its products.

With two years to the next Asama races in 1959, it was decided to send the leading factory rider, Fumio Ito, to take part in the international North American Catalina Island meeting during the summer of 1958. The decision was based of Yamaha's view that the United States represented the most important export outlet for its bikes. Consequently, Yamaha became the first of the Japanese companies to venture into this most important sphere. Ito eventually finished sixth after falling in the early stages and wetting a plug. Even so, the Yamaha management team were pleased with the result, and it was to influence future company policy in a very important way. What happened was that the Yamaha race shop was split into two separate factions. One was to develop specialised machinery with which to take part in European and the FIM World Championship Grand Prix series; the other was to develop the company's YDS-based production racer to take advantage of the AMA (American Motorcycle Association) competition rules, which outlawed full GP bikes.

Again, Yamaha had made the first move; others would follow later. And yet when it came to actually entering the European GP scene, of the 'Big Three' Yamaha ended up being a distant third. Honda was the first in 1959, followed by Suzuki in 1960, and Yamaha a year later. Why? One reason, was a desire not to be rushed and to follow the policy of 'quality first' at all costs. Another was the row that had developed at the 1959 Asama races when Yamaha had withdrawn in protest at Honda's new 250 four-cylinder model. Probably the major reason, however, was the success gained by the East German MZ factory with its innovative disc-valve two-strokes in Europe during 1959-60.

Until the appearance of the new MZs, the 1950s had seen the lightweight GP scene dominated by four-strokes such as FB Mondial, NSU, and MV Agusta. DKW and Montesa had tried to match these with piston-port designs, but had failed, while Suzuki did very poorly until it received a short cut to success when Ernst Degner defected from MZ at the end of 1961.

But Yamaha had a trump card of its own. This centred around the Showa marque, which had built a 125 disc-valve racer for the 1959 Asama races and had then promptly been taken over by Yamaha. Following this, Yamaha started work on their own disc-valve machines with which they would make their debut in 1961. This came at Clermont-Ferrand where, for the French GP, Yamaha fielded a team comprising Ito (who had campaigned a German BMW Rennsport flat-twin in the 500 cc World Championship with a fiery riding style in 1960), Noguchi, Makuko, Oishi and Sunako.

In the event, Noguchi came home seventh in the 125 race, while Ito finished in the same position in the 250 cc class. As for machinery, this comprised the RA91, a 123 cc (56 × 50 mm) single with twin carbs feeding the engine from both ends of the crankshaft, and the RD48, a development of the short-stroke 246 cc (56 × 50 mm) parallel-twin engine first used at Asama back in 1957, but with disc valve induction.

Power output figures for these first Yamaha GP machines were 20 bhp at 35 bhp, both at 10,000 rpm. In addition, both had six-speed, close-ratio gearboxes and dry clutches. Lubrication was a combined affair – petroil mixture, with extra protection being offered by a pump providing a separate supply of oil to the main bearings. The pump was mounted on, and driven by, the gearbox. Ignition was by magneto.

Compared with the Suzuki effort of the same period, Yamaha's 1961 GP performances, although showing the machines to be down on speed to the all-conquering Hondas, at least proved they were generally reliable – even in long races, such as the Isle of Man. Therefore, it remained a mystery to many why Yamaha did not return to Europe until the 1963 season. The reason was financial: having committed itself to what proved to be a disastrous programme of moped and scooter development during 1960 and

YAMAHA for 1973

With a variety of models ranging from 50 cc to 750 cc, YAMAHA, once again, leads the motorcycle world. These models are all designed for individual tastes and performance pleasure. The light- and middleweight machines are available with responsive, 2-stroke engines while the top-of-the-line heavyweights are powered by husky, 4-stroke engines. YAMAHA motorcycles, either 2-stroke or 4-stroke, are all characterized by performance, quality, ease of operation, and safety.

Avec une gamme de modèles alliant de 50 cc. à 750 cc., YAMAHA, une fois de plus, est en tête du monde motocycliste. Ces modèles sont tous conçus pour satisfaire aux goûts de chacun et apporter la joie de la performance. Les machines légères et moyennes sont propulsées par de nerveux moteurs 2-temps tandis que les machines les plus lourdes sont équipées de robustes moteurs 4-temps. Les motocyclettes YAMAHA, qu' elles soient 2-temps ou 4-temps, sont toutes caractérisées par leurs performances, leur qualité, leur aisance de pilotage et leur sécurité.

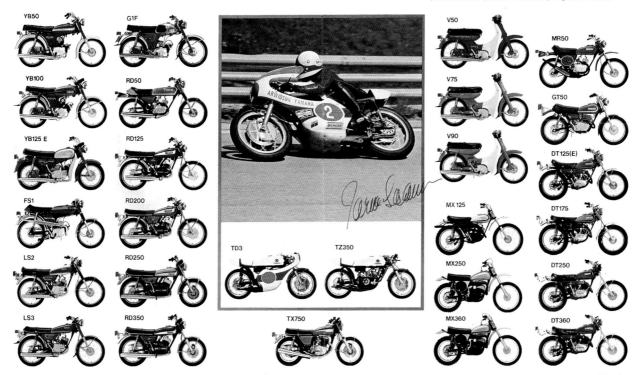

By 1973 Yamaha meant two- and four-stroke models on the street, and TD/TR (TZ in some markets) road racing machines. Until he was killed at Monza that May, the Finnish star Jarno Saarinen looked all set to scoop the 350 and possibly the 500 cc world titles. With his death the Japanese factory had to wait a little longer for success

1961, Yamaha simply could not afford a costly European racing venture in 1962. Obviously, this was a considerable setback to the racing department, and it is to their credit that they were to come back in 1963 to challenge Honda and Suzuki so effectively.

This lack of finance wasn't quite so important with the standard production machines, and Yamaha was still able to offer some excellent bikes in this period, most notably the YDS series of sporting two-fifties.

As the decade progressed so the range of standard production models increased, with North America as a particularly important market. Among the most notable of these was the YL1 (100 cc twin), YM1 (305 cc), YDS3, Ascot (Street Scrambler) and YG1 (80 cc), plus the TD series of 'customer' racers. But it was to be the 1970s that really established Yamaha as a truly significant force in the motorcycle world . . .

Lightweights

Like the rest of the 'Big Four' Japanese manufacturers, Yamaha have made it a priority to offer as wide a range of learner bikes as possible. Even though these ultra-lightweight motorcycles, scooters and mopeds rarely receive much attention from the press boys, they nonetheless play an important part in the campany's two wheel activities. Brand loyalty is created at an early stage in the rider's career; if you only offer big bikes you have to work even harder as many riders get attached to their early machines. Lightweights comprise, commuter bikes, learner bikes, scooters, step-thru's and mopeds.

Probably the longest running Yamaha lightweight is the YB100, which began as the YG1 with an 80 cc engine in the early 1960s and is still available today. The secret of its success stems from good design. The 97 cc (52 × 45.6 mm) pumps out almost 10 bhp from its single-cylinder rotary valve induction engine, which features pump lubrication. Today both these features are common in two-strokes, but three decades ago they were at the very forefront of technical innovation.

Designed very much with the short journey commuter in mind, the YB100 has deeply valenced mudguards to ward off road splashes, a fully enclosed chain guard to keep lubricant off the rider's clothes and rocker-pedal gearchanging to keep the shine on his or her shoes. Other features include a low 785 mm seat height, drum brakes, four-speed gearbox and efficient silencer. The result is a practical, durable machine.

Another long running model is the famous FS1E – the 'Sixteener Special' which over the years has been responsible for providing the first wheels for countless youngsters. Like the YB100, the FS1E employs rotary valve induction for its 49 cc (40 × 39.7 mm) engine. Even though performance is now limited to 3 bhp for most markets the machine is a true motorcycle. It looks like one, rides like one and is still attractive after all these years. The bike is fitted with seventeen-inch wheels, genuine motorcycle suspension, a four-speed gearbox with manual clutch to teach you how to shift gears correctly, a reasonable 780 mm seat height and a 9 litre fuel tank right where it should be, between the rider's knees.

Today Yamaha produce it in either drum brake (FS1) or disc front/drum rear brake (FS1DX) form, but some markets only have one version. In the UK early examples (up to 1977) were capable of speeds up to almost 50 mph, but current British laws restrict maximum speed to 30 mph. Many of the original 1972 models are still running – quite some achievement!

Two other excellent Yamaha lightweights from the 1970s were the RS100 (97 cc – 52 × 45.6 mm) and RS125 (123 cc – 56 × 50 mm). Both

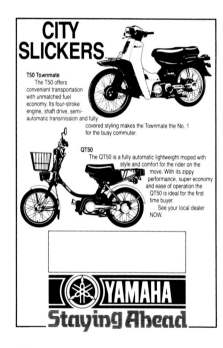

Above
Dealer advertisement for two models from the lightweight commuter range, the T50 Townmate and QT50. The Townmate had a four-stroke engine, shaft drive and semi-automatic transmission; the QT was a fully automatic moped with an horizontal two-stroke power unit

these two-stroke singles feature seven-port barrels and torque induction
(reed valves) together with pump lubrication and five-speed gearboxes.
The formula is survived today by the RXS100 with a 98 cc (50 × 50 mm)
five-speed motor. Its Yamaha Energy Induction System (YEIS) is a remote
storage chamber linked to the inlet tract which holds incoming fuel when
the engine is on its exhaust stroke and the reed-valves are closed. When
they open again, the stored fuel rushes out from the YEIS chamber at a
higher velocity and volume than the normal charge. The result is greater
mid-range torque and a 10 per cent increase in fuel economy.

By the early 1980s the company had decided to go very much for the
'soft' option – the female customer. The result was a trio of entirely new
machines: two scooters and a moped. There was the Salient (49 cc –
40 × 39.2 mm) and Beluga (79 cc – 49 × 42 mm) scooters and the Passola
(49 cc – 40 × 39.2 mm) moped. All employed Torque induction, two-stroke
engines and automatic transmissions.

Introduced in the 1970s, the V50 and V80 were very much a copy of the
hugely successful Honda Super Cub Step-thru, except for the two-stroke
reed valve engines. The smaller moped came with a 49 cc (40 × 39.7 mm)
engine and two-speed automatic gearbox (with dual range manual
selection), while the V80 had a 79 cc (47 × 45.6 mm) motor and a three-
speed gearbox with centrifugal clutch. Both had pump lubrication and
chain final drive.

Following in the footsteps of the single-cylinder four-stroke SR500
(Chapter 9), came the SR250 (239 cc – 73.5 × 56.5 mm) and the SR125

(124 cc – 57 × 48.8 mm). Both were to follow their larger brother's cylinder head layout with their chain driven single overhead camshafts, but otherwise were quite different machines, the two smaller SRs being commuter specials. The 125 in particular proved to be an excellent, if unexciting, little bike with economy, simplicity, safety and rider comfort as its priorities. Its dual seat is the most comfortable found on any machine of this engine size.

The RD LC range was extended downwards in the early 1980s to include the RD125LC and RD80LC (the RD50 was air-cooled). Both liquid-cooled models featured single-cylinder motors with reed- valve induction, electronic ignition, six-speed gearboxes, cast alloy wheels, bikini fairings, monoshock chassis and hydraulically operated disc front brakes. The 79 cc (49 × 42 mm) RD80LC produced 10 bhp at 6500 rpm, while the RD125LC was good for 21 bhp at 9500 rpm (restricted to 12 bhp in some

Above
New for 1988, the TZR125 was initially available with the choice of fairing at extra cost. Naked example allows full view of the 124 cc (56.4 × 50 mm) 2-stroke single-cylinder engine with its liquid-cooling, crankcase reed-valve induction and six-speed gearbox

Above

The 1993 TZR125 with the more sedate SR125 four-stroke ohc single.
Sandwiched between the two is a pair of the new-for- 1993 TDR125s

countries, including Britain) and over 80 mph. In 1985 the T80 (Townmate) replaced the V80. Main differences were a switch to a four-stroke engine and shaft final drive.

The following year the T50 arrived, simply a miniature version of the larger machine. Economy in both engine sizes was considerably improved and the shaft final drive was another sales boon, which even Honda couldn't match.

Also notable at this time was the QT50, a moped aimed very much at the economy-conscious rider.

But the really important news was the new liquid-cooled TZR125 (and its trail bike brother the new liquid-cooled DT125), which arrived in January 1987.

Power was provided by a new 124 cc (56.4×50 mm) crankcase reed-valve torque induction, two-stroke single, with YPVS (Power Valve). Crankcase reed-valve induction, as used on many of the factory's racing machines, significantly increases inlet efficiency and improves engine performance at all rpm. Featuring fibre reeds, acceleration and throttle response were much improved, and high-rpm reed surge eliminated. Virtually half a TZR250 engine, it had the same porting and exhaust pipe specifications.

To increase low and mid-range power the round slide VM26SS Mikuni carb, again like the TZR250, was equipped with a power-jet system. Mid-range running was much improved, fuel being sent to the engine in a more efficient way, instead of relying solely on the main jet/slide as previously.

A Deltabox frame was employed. This was fabricated from sheet steel using a 1.6 mm-thick outer wall and a 1.2 mm inner wall. The offside downtube being removable to facilitate engine servicing. While the frame was about the same weight as the now discontinued RD125LC, torsional rigidity was increased by no less than 70 per cent!

The front forks came directly from the RD125LC. Featuring 33 mm diameter stanchions and a fork brace, these contributed significantly to the machines excellent stability at all speeds.

Fixed-rate Monocross rear suspension featured a box-section steel swinging arm, controlled by a De Carbon-type gas/oil shock absorber, and pivoted on needle roller bearings.

A 16 inch, cast alloy wheel sported a slotted 245 mm disc operated by a twin-pot opposed-piston caliper, while the 18 inch rear wheel incorporated a drum brake.

The fuel tank held 12 litres (2.5 gal) of petrol with the airbox mounted underneath the tank's rear section. Like the TZR250, a flush-fitting aircraft-type aluminium tank cap was used.

Other details included clip-on handlebars and the instrumentation comprised a large tacho and a smaller speedo that sandwiched the water temperature gauge, together with a main switch that incorporated a steering lock.

Two versions were initially offered: the TZR125F, complete with full fairing, and the unfaired TZR125 – the latter being a cheaper option. Again, like the RD125LC, both versions of the newcomer were available in either full-performance or restricted (12 bhp) learner guise. A mini update for the 1991 season included the substitution of the drum rear brake with a 210 mm disc, a 17 instead of 16 inch front wheel and new graphics.

The TZR125 has proved itself one of the most successful Yamahas of recent years. I know of many examples that have survived unmerciful abuse from their inexperienced owners, when many lesser machines would have expired long before – this is one tough little bike and a credit to Yamaha's engineering practices and expertise.

1993 SR125 A3 model. The 4-stroke, sohc single-cylinder engine has a capacity of 124 cc (57 × 48.8 mm). Maximum power is 12 bhp at 8500 rpm. Other features include maintenance-free CD ignition, five-speed gearbox, an electric starter, spoked wheels, front disc brake and probably the most comfortable dual seat on any motorcycle of this size

Road Racing

As outlined in Chapter One, lack of capital was the reason Yamaha didn't mount a serious Grand Prix challenge until 1963, and when it did the company concentrated its efforts in the 250 cc class in the shape of the new RD56 model (although the company did build a prototype 125 single, based on half the RD56 unit). The engine of the RD56 was similar to the earlier RD48, but intense development had boosted maximum power by 10 bhp to 45 bhp at the higher engine speed of 11,000 rpm. Another important change concerned the frame. Yamaha engineers reasoned that it was not just a case of extracting maximum power, but also a matter of ensuring that the rider was able to make full use of the available horsepower – something which Honda could have learned from! The new chassis bore a remarkable resemblance to the British Norton Featherbed, and its handling was the best of any Japanese machine of the era. However, the biggest talking point of the new bike was the seven-speed gearbox, making it the first racer in history to be so equipped.

After an 18-month absence from European circuits, the Japanese factory's first port of call was the Isle of Man, where the team booked into the Douglas Bay Hotel during mid May. Four riders had been entered for the following month's 250 cc TT: Fumio Ito, Hiroshi Hasegawa, Yoshikazu Sunako and the company's first foreign signing, Tony Godfrey. The latter was a rider of considerable ability, having won the 500 cc ACU Star (British Championship) in 1958. During practice, Ito, Godfrey and Sunako were third, fourth and fifth fastest respectively, proving that the development team had done its homework well.

In the race itself, both Ito and Godfrey surprised more than a few people by leading the field at the end of the first lap. Shortly afterwards however, Godfrey stopped at Kirkmichael to change plugs, which took some two minutes, and later crashed attempting to make up lost time. He was subsequently rushed to hospital by helicopter, suffering from severe head injuries. Ito went on to finish second, a mere 27 seconds behind the race winner, Jim Redman (Honda four), at an average speed of 94.55 mph. Many felt that if Ito had not taken a very long 53 seconds for his pit stop he would have won! Team-mate Hasegawa came home fourth. Except for poor Godfrey's injuries, it had been an excellent comeback for the company.

Very rare 1964 246 cc (56 × 50 mm) TD1A production racer at Silverstone in September that year, rider D J Page. These early over-the-counter 'customer' racers were neither fast nor reliable enough to be regular winners

Above

Mick Grant rode Padgett Yamaha's in the 1972 Isle of Man TT. The Yorkshireman is seen here on the Padgett TD3 descending Bray Hill, Douglas. The TD3 (and TR3) were new that year. Their most notable innovation was the six-speed transmission. In addition, once again, there were power increases, with the 250 pumping out 49 bhp and the larger TR3 58 bhp

Right

World Champion Phil Read on the works 250 Yamaha V-four he rode to second place in the 1967 Isle of Man TT. Coded RD05A, the engine displaced 246.3 cc (44 × 40.5 mm), pumped out 60 bhp at 14,000 rpm and was good for nearly 150 mph

Yamaha had already decided only to contest the TT, plus the Dutch and Belgian classics. In Holland Ito gained another second, and then went one better by gaining the factory's first ever GP victory over the ultra-fast Spa-Francorchamps circuit, trouncing the mighty Honda team in the process. Ito's team mate, Sunako, took second ahead of Provini on the super-fast Morini single. Hondas took the next four places.

After this sensational result, many assumed that the team would stay in Europe to contest the remaining rounds, and Yamaha's team manager, Hiroshi Naitoh, telephoned his bosses in Japan to request that his team be allowed to stay on. All hopes of this happening were dashed, however, when shortly afterwards the European press received a briefly worded cable:

'Not participating further this year – Yamaha Motor'.

The final round in the 1963 World Championship series was held at Suzuka – the first time that Japan had been granted classic status. Yamaha's management lost no time in authorising the preparation of its machines for battle. With Godfrey sidelined, the team decided to bring in another British rider to augment Ito. This was Phil Read, who had already made a name for himself on private Nortons and as a works Gilera rider with Scuderia Duke that summer.

In the 250 cc race at Suzuka, the Englishman led for most of the race, before finally having to settle for third spot after experiencing plug trouble. Yamaha realised that here was a future world champion, and promptly signed the Luton rider for the 1964 season. Ito rode his usual excellent race to finish second and earn third place in the 1963 championship series – not bad for only four rides!

Above

The Yamaha production racers really took off in 1969 with the TD2 and TR2; next came the TD2B and TR2B (1970); TD3 and TR3 (1972); and finally the watercooled TZ models in 1973

Right
Another Padgett sponsored Yamaha rider was Peter McKindlay, seen here on an overbored TZ350, practising for the Senior TT in June 1975. Originally from Stourbridge, Worcestershire, Peter had just gained his first GP points, with sixth place in the 500 cc French Grand Prix, but was killed practising for the F750TT when he crashed his TZ700 four at Milntown Cottage. His death was a great blow to the whole Padgett family, as Clive Padgett commented recently, 'Peter worked and lived with us, so we felt his passing very deeply'

The 1964 GP season got under way at Daytona, USA, in February – the first time the Americans had run an FIM-sanctioned GP; Yamaha sent Ito and Read, but both were destined to retire – a wasted journey.

Next came the Spanish GP at Montjuich Park, Barcelona. Here, Read on the lone Yamaha (Ito had been injured in the non-championship Malaysian GP in March), finished third. The following week, the GP circus journeyed north to Clermont-Ferrand for the French round. Read, still riding the 'old' 1963 RD56, won a hotly contested race over the tortuous five-mile circuit. This victory meant that the Yamaha star was leading the championship series.

The TT that followed produced a few surprises in the Yamaha camp – even before the racing started. Tony Godfrey, who had recovered from his serious accident of a year before, declined to take up his entry. Yamaha then handed Godfrey's machine to Ulsterman Tommy Robb, who had just been sacked by rivals Honda. Ito was in the Isle of Man, but never actually rode. All through practice, he had assured the world's press that he would be 'out in the next session', but it never happened. Although recovered from the concussion and shoulder injury sustained in Malaysia, Ito was destined never to race again – very much an anti-climax for a man who had played such an important role in the Yamaha racing effort.

News was also released of an entirely new 125 cc class machine. This was a twin, not a rehash of the previous year's prototype single. Finally, Yamaha had pulled out all the stops for the Isle of Man and produced an uprated version of the RD56. This sported modified streamlining –

developed with the aid of a wind tunnel – and its twin-cylinder, air-cooled two-stroke engine was claimed to produce 10 per cent more power at almost 50 bhp. Eight of the new bikes were in the Island, comprising a practice and race mount for each team member, the fourth man being another new signing, Canadian Mike Duff. All this effort was to no avail, however, because not one Yamaha finished the race. The only consolation was setting the fastest 250 cc lap – Read at 99.52 mph.

Highlights of the Dutch TT that followed were two great duels between Read and Honda-mounted Jim Redman. In the 250 cc race, the pair fought out a fantastic battle, which Redman finally won after the Honda team got to within a second of Mike Hailwood's 500 cc MV absolute lap record.

The second Read-Redman dice came in the 125 cc event. Riding one of the brand new RA97s, Read surprised everyone by giving Redman, who was on one of the four-cylinder RC146 models, a terrific run for his money. The RA97 was virtually an RD56 in miniature; its specification included a bore and stroke of 44×41 mm, an eight-speed gearbox and 28 bhp at 13,000 rpm.

There was no 125 cc class at the Belgian GP a week later. Here, having his third outing on a works Yamaha, Mike Duff not only became the first Canadian to win a classic, but also raised the nine-lap race record for the ultra-fast eight-mile circuit to 118.41 mph. In the process, Duff finished more that half a minute ahead of the reigning 250 cc World Champion, Redman, who was second on his Honda. Earlier, Read had just snatched the lead from Redman when his RD56 seized as they sped through the flat-out 140 mph swerve in the middle of the downhill Masta Straight! Robb was fourth behind the MZ of Alan Shepherd.

In mid July, at the West German GP, Read and Duff found their smaller twins no match for the Hondas and Suzukis; Read slid off and Duff went out with mechanical trouble. But Read then began a series of 250 cc victories that were to result in his and Yamaha's first ever world title – and the first achieved by a two-stroke in the 250 cc category.

With three straight wins for Yamaha at Solitude, Sachsenring and Dundrod, Honda panicked and brought out a 250 six-cylinder machine for Redman to challenge Read at Monza. This tactic failed, however, and Read won again to take the title. Even though Redman won the final round, the Japanese GP at Suzuka, on the new six, there were only five finishers, and Read was not among them. Many observers predicted a poor year for Yamaha in 1965, but they were proved wrong, with Read retaining the

The great Mike Hailwood gets ready to push his Martini-Yamaha into life at the start of the 1978 Senior TT

Right

Typical privateer racer of the late 1970s was this 347 cc Spondon-framed TZ. Many riders opted for afermarket frames not only from Spondon, but from the likes of Harris, Bimota and Maxton as well

Left

South African Jon Ekerold won the 1980 350 cc world title riding a Bimota-Yamaha, and followed this by taking second place behind Kawasaki factory rider Anton Mang the following year. Ekerold is shown taking part in the 1981 Senior TT

Below left

Mac Hobson was one of the leading British sidecar aces to use the fearsome TZ700 engine. The 694 cc (64 × 54 mm) four Yamaha motor produced 97 bhp at 10,400 rpm. Gearbox was a six-speeder

championship with seven victories in the 13-round series. In addition, team-mate Duff ended the season in second spot. However, if Redman hadn't broken an arm at the first of the European rounds in West Germany, things might have been very different, as he showed by winning three GPs mid season on the trot, before being sidelined again following an accident in Ulster when he broke his collarbone.

Meanwhile, the 125 cc RA97 had been converted to water-cooling for 1965, but was only seen occasionally. Even so, Read won in the Isle of Man, while Duff took the chequered flag in Holland. Meanwhile, Yamaha's main effort went into its successful defence of the 250 cc title.

The TT had also seen the debut of Bill Ivy, whose top placing in his new team came at the final round in Japan, where the diminutive Kentish rider was placed third in the 250 cc race. For 1966, Ivy replaced Duff, and so began a period of intense competition for honours between Yamaha team-mates Read and Ivy. On the water-cooled RA97, Ivy won in Spain, the Isle of Man and Holland, and Read in Finland, but Taveri took the title for Honda with the new five-cylinder model.

In the larger class, Yamaha had responded to Honda's six-cylinder machine by introducing a brand-new V4 250 at Monza in September 1965. Even though, like the Honda the year before, it didn't win at its debut, the scene was set for its full scale introduction the following year. But Yamaha had not counted on Honda signing Mike Hailwood! This legendary rider went on to dominate the 1966 250 cc World Championship, winning ten of the 12 rounds. At the remaining two, a Bultaco took the Ulster, while

Above

Easter Anglo-American Match Race Series, 1981; Don Vesco works on Dave
Aldana's bike at Mallory Park. TZ750 (OW 31) produced a staggering 140 bhp at
10,700 rpm. Top speed was over 180 mph on optimum gearing

Left

American Kenny Roberts scored a hat trick of 500 cc world titles for Yamaha in
1978, 1979 and 1980. He is seen here on his way to second spot in the 1981 British
GP at Silverstone

Hasegawa beat Read in Japan. Yamaha was not helped by a series of big-
end failures and poor handling on the new V4.

September 1966 saw Yamaha switch to an all-V4 effort with the
introduction of a 125 model. Theoretically, this put the company in a strong
position for the 1967 championship. As for technical details, the 1967 125
V4 (RD05) generated 35 bhp at 16,000 rpm from its 124.6 cc (35 × 32.4
mm), 90-degree disc-valve engine and featured a nine-speed gearbox.
Maximum speed was almost 130 mph. The larger model (coded RD05A)
displaced 246.3 cc (44 × 40.5 mm), pumped out 60 bhp at 14,000 rpm and
was good for nearly 150 mph. 'Little Bill' Ivy, with eight wins in the 12-
race series, easily won the 1967 125 cc world title and proved himself a
brilliant rider – capable of matching and beating the best. Read, with two
victories, backed him in the runner-up position.

With Hailwood still Honda-mounted, things were much more difficult in

Left

During his championship years, Roberts' great threat came from Barry Sheene. But after being pipped for the title in 1978 and 1979 Sheene opted to switch from factory Suzuki to private Yamaha (backed by the Japanese Akai company) but it wasn't always a happy liaison. In 1981 Sheene finished 5th in the title chase. He is seen here in winning form at Oliver's Mount, Scarborough on his 498 cc TZR500 in the Motor Cycle News Superbike race, 6 September 1981

Above

Another Yamaha rider to feature in the World Championship series was the Italian star Gian Rossi, seen here at Silverstone in 1981; he finished fifth in the 1980 title chase

Above

Kenny Roberts at Daytona in 1982. The 'King' went on to become a team manager after his racing days were over. He has stayed loyal to Yamaha not only during his racing career, but also with his Team Roberts over the span of two decades

Left

During the late 1970s Charlie Williams won many Isle of Man TTs on a variety of Yamahas, including the 500 class. He is pictured here at Union Mills on a Mitsui-backed TZ500 four

the 250 cc class, but as proof of the progress made by the Yamaha development team, the 1967 race results were much improved from the previous year. In fact, Read finished the season with a larger number of basic championship points than Hailwood, but the Honda man took the title, having gained an extra victory (Honda five, Read four). In addition, Ivy scored two victories, as did Ralph Bryans for Honda – close indeed!

First Honda then Suzuki bowed out in 1968, leaving the field clear for Yamaha to clean up in the 125 and 250 cc championship series. The only rivalry was between team-mates Read and Ivy! For the 1968 season, the smaller V4 was giving 42 bhp at 17,000 rpm (135 mph), and the 250 70 bhp at 14,400 rpm (155 mph). With these power output figures available, both Honda and Suzuki would have found the going tough had they remained on the scene. As it was, Yamaha machinery totally dominated the 125 and 250 cc categories that year.

Above

Englishman Alan Carter became the youngest ever rider to win a Grand Prix, when he rode his Mitsui TZ250 to victory at Le Mans in May 1983

Left

Dave Davis, the 1982 500 cc Auto 66 Champion, on a Maxton framed overbored TZ twin at Elvington, North Yorkshire that year

According to Yamaha factory orders, Read was supposed to win the 1968 125 cc title, which he duly did, while Ivy was down for the 250 cc. However, Read had no intention of following these orders and disregarded them as soon as he was sure of taking the 125 cc class, so the two riders clashed in a bitter fight to the finish. Read, who had lost 'his' title to Hailwood in 1967 because of the final ruling, wanted to regain the title at all costs.

At the end of the ten-round series, Read and Ivy were neck and neck with the same number of points, each having five victories. It couldn't have been closer, so the FIM decided to award the title based on the four races where they were both first and second – Holland, East Germany, Czechoslovakia and Italy. The difference was a mere 2 minutes, 5.3 seconds in favour of Read. All this created a distinct air of hostility between the two parties involved, although it is unlikely to have been a factor in Yamaha's decision to follow Honda and Suzuki and quit at the end of 1968.

The Read-Ivy affair seems strange, especially knowing that Read had

Above

'King Kenny' — Kenny Roberts, America's (and Yamaha's) first ever 500 cc World Champion; this fine study was taken in 1983

Left

Kenny Roberts in full flow on a 500 cc Yamaha during the 1982 World Championship season

backed Ivy being signed to Yamaha in the first place. Another faction within the team had wanted John Cooper instead! One also has to remember that the two men were quite different. Read was only interested in winning races, and to many others he seemed 'distant'. Bill Ivy, on the other hand, was someone who lived life to the full and didn't take things too seriously — except for that 1968 250 cc title run-in! Yes, things ended on a bitter note, but there is no denying that both Read and Ivy were superb riders; the problem was that each wanted the same title . . .

At the end of 1968, Yamaha pulled the plug on its full works efforts. The main reason for this decision was that its machines were about to be outlawed by new rules governing cylinders, gears and weight. Under the FIM's revised criteria for Grand Prix racing machines, the 50 cc class was to be limited to a single cylinder, six gears and 60 kg weight; the 125 cc category to two cylinders, six gears and 75 kg; and the 250 cc class to two cylinders, six gears and 100-plus kg. The new regulations were to be applied to the 50 cc machines in 1969, and to the others in 1970.

Above

Scottish rider Eric McFarlane, with his Team Ecosse Formula 2 machine at Donington Park 1984. The frame was in plate-alloy and was designed and constructed by Mike Harrison of Selston, Nottinghamshire. Two bikes were built for McFarlane to contest the World Formula 2 championship. The power unit had to be roadster based, so an LC engine was used. It lapped the Isle of Man TT circuit at 106 mph

Left

Carter again, this time on a Marlboro-Yamaha at Donington Park in the summer of 1982. Unfortunately, although he became a multi-British champion he never went on to the Grand Prix stardom many had expected during his early days. He finally quit racing at the end of 1992

After eight years of effort to catch up Honda and Suzuki, Yamaha was less than pleased and, as a result, decided to concentrate its efforts on the new TD2/TR2 twin-cylinder production racers.

This heralded a new period of glory, when works' supported riders took four 250 cc world titles (1970, 1971, 1972 and 1973) and three 350 cc titles (1974, 1975 and 1977). Yamaha twins also captured two titles in the 125 cc class (1973 and 1974). In the 500 cc class, Yamaha only gained one title (1973) before the appearance of American superstar Kenny Roberts in the late 1970s. Roberts took the 'blue riband' class for Yamaha in 1978, 1979 and 1980. He was also America's first road-racing world champion, creating

Above
'Steady Eddie' Lawson was the
mainstay of Yamaha's 500 cc world
championship success for much of the
1980s. He was the true professional –
reliable, smooth, fast and most
important of all, a consistent race
winner at the highest level

Right
The view most of the opposition got,
most of the time, of the combination of
Lawson and Yamaha

The combination of Steve Abbot and Shaun Smith was probably the best British sidecar pairing next to Steve Webster and Brian Hewitt; they are here during 1984 with the Team Ham-Yam machine at Donington Park

a path with grit and talent which many others were to follow.

Other notable Yamaha riders of the 1970s included Rodney Gould, Jarno Saarinen, Dieter Braun, Giacomo Agostini, Johnny Cecotto, Kent Anderson and Takazumi Katayama – plus our old friend Phil Read. All these won world titles with Yamaha during the decade, and all but Read received official backing for their efforts.

Like rivals Honda and Suzuki, Yamaha returned to a policy of full-scale works support for the 1980s that was to place the company firmly back on top, at least in the all-important 500 cc GP category.

The company's first success after Roberts' 1980 title came in 1984 when Eddie Lawson scored his inaugural championship success with victories in South Africa, Spain, Austria and Sweden to score 142 points, ahead of runner up Randy Mamola (Honda) with 111 points.

Lawson campaigned a 498 cc (56 × 50.6 mm) liquid-cooled, four-cylinder two-stroke. Coded OW76, this sported crankcase reed-valve induction, with five transfer and one exhaust port. Each piston used only a single ring and there were roller big-ends and needle roller small ends. A pair of pressed-up crankshafts sported four roller bearings per assembly. The dry clutch had five friction and six steel plates; gearbox was a six-speeder. Carburation came from a quartet of 34 or 35 mm flat slide Mikunis, depending upon the circuit.

As for the frame, this was a twin spar aluminium device with taper roller head bearings and a box section aluminium swinging arm with needle pivot roller bearings; the lower end linked rear suspension had a single Öhlin shock absorber with a coil spring plus nitrogen pressure. Other details of the 1984 OW76 included Campagnolo wheels, Dunlop tyres, Brembo brakes and Castrol oil. It is also worth mentioning that although Kenny Roberts was no longer to be seen on the track in Grand Prix events, he remained a force in the sport as manager of the official 1984 Marlboro-Yamaha team in the 250 cc class with Alan Carter and Wayne Rainey as riders. The subsequent fortunes of their competitors is really the story of GP racing from that time onwards – British decline and American dominance.

Lawson went on to win again in 1986 and 1988, before leaving to join Honda, and ultimately the Italian Cagiva concern. In 1990, Wayne Rainey began a three year hold on the championship crown, all this coming after his first 500 cc Grand Prix victory at the British GP at Donington Park back in August 1988.

Before leaving the 500 cc class it is worth recalling that from the 1992 season onwards, Yamaha has provided replicas of its factory engines for privateers to fit into frames such as the popular British Harris. All this came as a response to the ever-shrinking grids for the class at Grand Prix level, due to the lack of competitive, non-works machinery. For this Yamaha should be thanked even though in Britain an engine alone costs more than £75,000.

Providing 'customer' racers has long been a Yamaha hallmark, going back as far as the original 250 TD1 of 1962. Then came the TD1A (1963 and 1964); TD1B (1965 and 1966); TD1C (1967 and 1968); TD2 (1969 and 1970); TD2B (1971); TD3 (1972 and 1973); and the liquid-cooled TZ250 (still available in 1993) with first parallel cylinders, then reverse cylinders (1988, 1989 and 1990); and finally the GP-based V-twin concept from 1991 onwards.

For many years Yamaha has also offered a 350 over-the-counter racer. The first of these was the 348 cc (61 × 59.6 mm) TR2 of 1969 and 1970. Next came the TR2B (1971) and TR3 (1972), before becoming the TZ350 with liquid cooling, which ran with regular updates until production ceased in the early 1980s (at the same time as the class was deleted from the World Championship series).

Countless club, national and even international races have been won by these machines, which together with the success garnered by Yamaha at works level has ensured a unique place for the marque in the annals of motorcycle racing.

Above

Giacomo Agostini, the former MV Agusta and Yamaha racing legend. During the 1980s he became the Marlboro-Yamaha team manager

Left

The 1986 Marlboro-Yamaha team. Left to right, Eddie Lawson, Giacomo Agostini (manager) and Rob McElnea

Above

The 1988 version of the long-running liquid-cooled, twin-cylinder TZ250 over-the-counter racer. The 249 cc (56 × 50.7 mm) engine produced 70 bhp. There were twin 36 mm Mikuni carbs, CDI ignition, 8:1 compression ratio and a six-speed 'box

Left

Wayne Rainey on his way to his first ever GP victory; the 1988 500 cc British event at Donington Park in August that year. The factory YZR500 was a liquid cooled V4 two-stroke; its 499 cc engine generating over 148 bhp and capable of 180 mph. Other features included YPVS and crankcase reed valve induction

Above

The Japanese Yamaha works rider
Takazuma Katayama before the start of
the 1978 Senior TT; he finished fifth in
the 500 cc world championship series
that year

Right

1992 model TZ250 V-twin at Brands
Hatch, 18 October 1992; rider is Mick
Otter from Rotherham, South Yorkshire

RD – race developed

The first RD models appeared in 1973 and each was to set new standards of all round ability for a road going two-stroke twin in its respective capacity class. It was also to establish Yamaha as a maker of 'sports' rather than pure street bikes, something the company has managed to preserve to this day.

This quartet of 1973 trend-setters were the RD125 (124 cc – 43×43 mm), RD200 (196 cc – 52×46 mm), RD250 (247 cc – 54×54 mm) and RD350 (347 cc – 64×54 mm).

These bikes are also very important in any history of Yamaha as they 'carried' the company through the mid-1970s following the patchy success achieved by a number of four-stroke parallel twins.

Before the introduction of the new RD-series engines in 1973, all the company's piston-controlled two-stroke twins had been something of a compromise. This engine type suffered from the contradictory tasks the piston had to perform. The need for speed meant that piston-port designs tended to suffer from narrow power bands, making them difficult to ride, if not downright unreliable in operation.

The reed-valve appeared the ideal solution to the problem. In its simplest guise, a reed-valve is purely a uni-directional device, opening as a vacuum is created under the rising piston, allowing new charge into the engine and

Left

The first sporting two-fifty roadster from Yamaha was the YDS1 (1959-62); then came the YDS2 (1962-64); followed by the YDS3 seen here which ran from late 1964 until mid 1967. The 246 cc piston ported parallel twin engine had a capacity of 246 cc (56×50 mm) and the maximum power output of 28 bhp was produced at 8000 rpm. The YDS3 was notable as the first Yamaha sports model to feature Autolube (pump) lubrication

closing as the piston descends, preventing the charge from leaving the way it entered. As described in one of my earlier books, Classic German Racing Motorcycles (Osprey Publishing, 1990), a system had been pioneered by the German DKW marque during the inter-war years, coupled with piston windows and deflector crowns.

In Yamaha's reed-valve (promoted by the company under the title 'Torque Induction'), the system was able to make use of modern two-stroke technology to great effect. Each reed-valve assembly consisted of a wedge shaped aluminium die-cast block, with a flat base plate. This base plate bolted on to the barrel, with the wedge pointing down into the cylinder. To each block were bolted four pairs of metal strips, two adjacent on each side of the wedge. The two stainless steel strips consisted of an outer curved thick strip of steel clamped to the reed block, sandwiching a very thin flexible strip of steel in between. The aluminium block was cut

away under the thin strip so that it was clamped at one end, supported by the point of the wedge at the other, with an opening within. The outer thick strip acted as a stop, controlling both the reed's travel and curvature. To provide some degree of cushioning as the valve flapped back and forth, the reed-case was coated in a micro-layer of neoprene, a synthetic rubber compound.

Notwithstanding its German origins, Yamaha's reed-valve was truly innovative and should be explained within the context of the company's other two-stroke technical developments.

During its Grand Prix campaign in the sixties, Yamaha enthusiastically pursued the aim of a perfectly scavenged engine. First there had been the use of 5-port cylinders, in which different directions of auxiliary and main transfers had significantly improved exhaust gas extraction. However, there was still an area of cylinder wall that made no contribution to the scavenging process, namely above the inlet port. This section of a wall on an engine with disc valve induction was liberally sown with transfer ports, all of which contributed to the extraction of exhaust gases. With the introduction of the reed- valve, Yamaha saw an ideal opportunity to put this to good use.

A cross section of the inlet port now looked like an inverted cross section of a mushroom, with a gully extending up to the same height as the other transfer ports. The cylinders were described as 7-port, counted as inlet, outlet, two main transfers, two auxiliary transfers and the extra port above the inlet. It all worked through the ingenious utilisation of crankcase and

Right

The first RD250 appeared in 1973. The 'square' bore and stroke dimensions of 54 × 54 mm (247 cc) were inherited from the DS7 (1970-1973). Likewise, the RD350 introduced at the same time shared the 347 cc (64 × 54 mm) sizes with the R5 which was produced in parallel with the DS7. This RD250 is competing in the 1974 Isle of Man TT

pressure changes. On the piston's rising stroke, the inlet port opened, and due to the vacuum in the crankcase, the reed snapped open, allowing the charge into the engine as the piston descended on its power stroke. The reeds initially closed as the crankcase was pressurised, but when the top of the piston cleared the top of the transfer port, the negative suction wave formed in the exhaust pipe was strong enough to crack the valve open and pull in some more charge directly from the inlet manifold. As the strength of the exhaust wave decreased, the valve re-closed to wait for the piston's ascent. Not only was the engine getting more charge, but more of the exhaust gases lurking at the back of the cylinder were expelled. To open the inlet port as early as possible, two windows were cut in the inlet skirt of the pistons. This exposed the valve to the crankcase vacuum, but still supported the body of the piston, and prevented it from rocking into violent self-destruction.

The new RD series were all equipped with reed valves. This not only meant a useful increase in performance, particularly in the mid range, but gave Yamaha the chance to trumpet the system under the guise of 'Torque Induction'.

It was probably the RD250 that benefited most with usable power now starting at 4000 rpm rather than 6000 rpm as on the piston port model it replaced – a full one third difference!

Production of the two larger RD models began in December 1972, with the 125 and 200 coming on stream two months later.

Actually, except for the reed-valve technology, little had changed in the

general design and styling between the RD250/350 and their predecessors, the DS7 and R5 (see Chapter One).

One major difference was the six-speed gearbox (five-speed on 125/200), primarily to enable the company to claim that its street bikes, from which the TZ racers were derived, had six gears, thus 'legalising' the TZ gearboxes.

The range of RD twins was to remain unchanged until the 1975 model year, when they all underwent a cosmetic update, together with a few technical changes that included giving the RD125/200 models disc front brakes (already enjoyed by the RD250/350 machines), and fitting the two larger models with new cylinder heads and revised gearbox ratios.

The following year the 350 was replaced by the 398 cc (62 × 64 mm) RD400. With an 8 mm longer stroke, the engine had almost square dimensions. In response to the additional power (up from 39 bhp to 40 bhp), not only were there new, stronger con-rods, but the complete crankshaft and crankcases were beefed up too.

In production racing, where they were most at home, the RD250 and 400 dominated their respective classes during much of the mid to late 1970s. The prestigious British Avon race series was won by Phil Mellor and Richard Stevens in the 500 and 250 categories respectively.

Throughout the rest of the decade the RD models were subjected to a number of detail improvements – mostly limited to the two larger versions. These included alloy wheels, thicker (35 mm) fork stanchions, disc rear brakes and electronic ignition. The latter first appeared in 1978

Right
This overbored RD400 – new capacity
430 cc – was built for drag racing by
Harry Barlow. The Leicester-based
two-stroke tuner ran Pro-Porting until
he left to work in the USA at the end of
1992. This machine represents the
pinnacle of air-cooled Yamaha two-
stroke development, at least of the
roadster based engine

with the model suffix 'E' and was in response to plug fouling problems
experienced by the larger twins. Despite some initial hiccups, the new
ignition was to finally eliminate the plug problem through a combination
of more precise timing and a fatter spark. Again, Yamaha set the standard
for future two-stroke practice.

The final variant of the air-cooled RD line was the 'F' model of 1979
aimed at countering the tough US EPA legislation. However, even this
was only a temporary postponement, with the result that the following
year Yamaha stopped selling two-strokes in the States, axed the RD range,
and in its place introduced the European-aimed RD250/350LC series
described in Chapter Six.

Thus died a family of models that had served Yamaha well, the RD
(race developed) air-cooled twins bowing out with pride.

DT – two-stroke trail

The all-round, on/off road bike was largely pioneered by the DT1 (Dirt Trail), a machine that made a fortune for Yamaha in the all-important American market.

In the late 1960s US distributors saw a demand for motorcycles on which the rider could take advantage of America's vast open spaces – customers wanted comfortable seats, long wheel bases and fat front tyres. The bikes also had to be reasonably lightweight and perform well enough for stateside riders to enter 'Sportsman' category events.

At first both the European and Japanese manufacturers balked at the idea, saying that it was madness to build an off-road bike unless it was a pure motocrosser or pukka trials iron. Gradually, the Europeans, most notably British and Spanish manufacturers such as Greeves, BSA, Bultaco and Montesa, came up with the type of bike the Americans wanted, albeit with an often truly spartan specification.

Then came Yamaha, and the DT1. As Cycle World observed in its February 1968 issue: 'The new DT1 is as American as a Coca Cola'. CW continued: 'The fact that it can be ridden on the street is an added bonus. Very efficient lighting and silencing make the DT1 a most pleasurable mount for going to market. Or, if called upon, to make a highway jaunt. The five-speed transmission allows 60 mph at a modest 6000 rpm, while

Left
The twin cylinder TDR250 used the same engine as the fire breathing TZR sportster and was marketed between 1988 and 1991, but never sold in the same numbers. The 249 cc (56.4 × 50 mm) engine was good for almost 50 bhp. Other features included liquid cooling, YPVS, crankcase reed valve induction, a six-speed gearbox, 18 inch front and 17 inch rear wheels rims and a dry weight of 137 kg (303 lb)

Right
By 1991 Yamaha was able to offer the excellent DT125R trail bike. Its liquid-cooled single cylinder 2-stroke engine featured Torque Induction, YEIS (Yamaha Energy Induction System), Autolube oil injection, a front disc brake and Monocross single shock rear suspension. Maximum speed was around 70 mph

bottom is low enough for plonking through the woods'.

Yamaha had produced what was probably the world's first truly dual purpose motorcycle; its innovative blend of performance, handling, comfort and practicality soon imitated by a host of other manufacturers – the 'trail bike' had arrived.

The DT1 was equipped with a single-cylinder two-stroke engine of 246 cc (70 × 64 mm). Output was around the 22 bhp mark, but a Genuine Yamaha Tuning (GYT) kit upped the power to over 30 bhp. The GYT comprised a single ring piston; a chrome-on-aluminium cylinder (replacing the standard iron-sleeved barrel); a racing expansion chamber; a central plug head for optimum combustion chamber shape; and a 30 mm (standard was 26 mm) bore carburettor.

Primarily a cosmetic update, the revised DT1-B of 1969 was the first model to appear in Europe in any numbers. From these early efforts came a whole family of Dirt Trail models ranging from the trail-moped class DT50 right through to the 397 cc (85 × 70 mm) DT400 offered between 1975-78. In among these came some excellent machines, most notably the air-cooled DT175 (1973-1985), and the liquid-cooled DT125, a bike that made its bow in the early 1980s and is still going strong a decade later.

Finally, a special mention should be made of the twin-cylinder TDR250 – a brave, but ultimately failed attempt at producing a trail-style bike with a pure street performance; it was certainly no slouch, as demonstrated by the 107 mph maximum speed and standing start quarter mile figures of 13.89 seconds/97.5 mph obtained by Performance Bike in 1988.

LC – headbangers choice

The world had its first glimpse of the new RD350LC in October 1979. Stealing the limelight at the Paris Show, the long awaited replacement for the air-cooled RD was here at last.

The original three-fifty was to herald the beginning of a whole new family of machines: the RD250LC, RD125LC, RD80LC, 350YPVS and ultimately the flagship RD500LC four.

In launching the LC series, Yamaha took several major technological steps, introducing cantilever suspension and liquid cooling to bring ordinary riders a new dimension in two-stroke performance.

The 250 and 350LC models were primarily aimed at Europe, and started to arrive in the spring of 1980. Compared to earlier Yamaha two-strokes they contained all manner of new developments. The most obvious was liquid cooling (LC). Yamaha used a mixture of 50 per cent distilled water and 50 per cent ethylene glycol (anti-freeze), forced around the aluminium cylinders by a crankshaft driven pump.

This basic arrangement had two distinct advantages; it minimised bore distortion – the cylinder having a more uniform temperature – and reduced the amount of mechanical noise. Strangely though, piston/cylinder clearance in the LC was actually greater than its forerunner, the RD.

Left
When the RD series of 2-stroke twins was finally axed in mid 1979 by American emission regulations, everyone thought that Yamaha would have finished with the sporting 2-stroke roadster. How wrong they were, as in 1980 came the 250 . . .

Right
. . . and 350 LC models. Effectively aimed at the European market, this duo of watercooled 'stroker twins set new standards of performance in their respective capacity classes

Commonality is a valuable feature of the LC series: 350LC barrels fit the 250LC; the 250 and 350LC barrels fit the air-cooled RD; and TZ barrels (up to 250G) fit the RD and LC. It would be logical to presume that the LC motor developed from the TZ; but it didn't! The result was a direct development from the RD series, bearing little resemblance to the piston-ported TZ racers of the same era.

A common factor of the RD/LC range was reed-valve induction; this system was explained at considerable length in Chapter Four, but it is worth mentioning that the arrangement was largely responsible for the LC's easy starting and lusty low/middle range power. Reed-valve induction was so good that Honda, Suzuki and Kawasaki had little choice but to follow suit, incorporating it in one form or another on their production machines. Staying with the subject of induction, the single-cylinder RD80LC and RD125LC were both equipped with the Yamaha Energy

Above

Many famous riders first made their mark in Pro-Am racing, like Kenny Irons (who later became a works rider for Yamaha), seen here on the start line at Donington Park

Induction System (YEIS). This simple device consisted of an empty bottle/reservoir connected to the inlet tract between the carb and reed block. Its operation was also simple, storing a small volume of petrol/air mixture normally compressed behind the reed block. Pressure waves were smoothed within the inlet port, making carburation cleaner and increasing the power developed at low speed. The 250 and 350LCs did not have the YEIS system as both bikes were 180 degree twins with large diameter balance pipes – the surplus charge building up behind the reed simply diverted to the adjacent cylinder.

Items such as the six-speed gearbox and crankshaft derived from the old RD models. Details may have varied but the basics were the same. Electrics was another area where much was as before. For example, the utterly dependable crankshaft-mounted generator/ignition needed no development, as from the late 1970s it had provided a valuable boost to the reliability factor.

What was new? Well, the cantilever suspension found on the original 250/350LC models developed from the YZ motocross machines. It offered increased rigidity and much longer rear wheel movement compared with the previous pivoting fork of the RDs with their twin shocks.

The single De Carbon-type rear suspension unit was tucked away under the frame and the unit received little in the way of cooling air – and what it did get had already been warmed by the engine and radiator. Not surprisingly, under arduous conditions the LC twins tended to suffer from damper failures.

Another major development on the LC series was the anti-vibration engine mounting system. Again simple, but effective, the motor was pivoted at the rear and retained by high deflection rubber bushes at the front. The engine simply bounced up and down; so noticeable is the effect that one can literally see the motor moving independently from the frame!

When first introduced the 250/350LCs gained a reputation for breaking exhaust pipes due to excessive engine movement. This was cured in 1981 when two additional rubber bushed tie-rods were added. These additional tie-rods, fixed to the underside of the crankcases and extending forward to the frame, allowed vertical but restricted lateral engine movement.

A common complaint on the air-cooled RD series was crankshaft failure due to lack of oil. This was the result of neglecting to check the oil level windows attached to the tank. Yamaha's response was to fit the 250/350LCs with 'idiot proof' panel mounted warning lights. The lamp, operated by a simple tank-mounted float switch, gave the rider ample warning of low oil level.

On the racing front the 250LC was a winner at its first attempt. This was in the production class of the 1980 Isle of Man TT. The 350 gained equal success by immediately dominating the 500 'proddie' racing class.

Between them the LCs won nearly everything in sight; from club
championships to major international production events. Attempting to list
their successes would fill the rest of this book. The simple fact was that a
road-going LC, with a modest amount of tuning, produced the sort of
performance that a few years previously would have been sufficient to win
a national championship.

By 'legal' modification of the LC engines for production racing it was
possible to extract some 46 bhp at 9000 rpm from the 250, and 54 bhp at
9500 rpm from the 350.

Formula 2 category racing was completely different; with TZ type
modifications allowed, the 350 LC engine produced around 75 bhp and
propelled the bike to almost 150 mph.

Graham McGregor won the 1984 Formula 2 TT at a record average
speed of 108.78 mph. His machine was an Arnold Fletcher tuned non-
power valve 350LC engine housed in a TZ frame.

Back in 1980 Yamaha introduced the power valve TZ250H racer; the
question then was only how long it would take the company to transfer
this device to a series production roadster. Sure enough, at the 1982
Cologne Show Yamaha unveiled the all-new, RD350LC, YPVS. At first
many observers thought that the 64×54 mm two-stroke twin was just a
restyled and uprated 350LC... how wrong they were. In fact it was
considerably different, the 350YPVS producing 59.1 bhp at 9000 rpm,
12 bhp more than the bike it replaced, and its chassis had a specification
that a few years earlier would have put any 350 racer to shame.

One of the important improvements over the original 350LC was the
YPVS (Yamaha Power Valve System). This was very simply a valve that
operated in the exhaust port window; the timing on a 9000 rpm, high
output engine would exceed 200 degrees to obtain the necessary port area
to handle the exhaust gas within the time available. This would produce,
in a conventional engine, a power delivery far worse than on the RD400;
as well as poor response at low rpm, the power would arrive like a kick
from a horse.

The answer was obviously to change the exhaust port height while the
engine was running, increasing the port height and area as the speed
increased. The Yamaha Power Valve system afforded high gas trapping
capability at low speeds and provided the necessary port area for peak
power at high rpm. Producing a mild mannered street racer was not the
only effect the YPVS system had; at low and medium rpm it also
restricted the amount of unburnt fuel that disappeared down the exhaust,
promoting more efficient running and therefore improved fuel economy.

The valve control system of the 350YPVS was electronic, but during
development the TZ mechanical centrifugal type (similar to a contact
breaker advance) was tried, as was a hydraulic system employing the

Above

The 1983 RD350LC YPVS model; it was the first production roadster from Yamaha to feature power valves

Right

A year later came the sensational four cylinder RD500LC. It produced just over 70 bhp and had a maximum speed of 148 mph

engine's water pump. The electrically driven system emerged a clear winner, consuming less power and having more accurate control over the valve than its rivals. The engine speed is first detected from the ignition and through a simple frequency to a voltage converter, the battery drove the 0-12 volt motor which in turn rotated the power valve: 0-5000 rpm = 0 volts (valve closed); 9000 rpm = 12 volts (valve fully open).

Since its introduction in its mechanically controlled form, the power valve has seen many manufacturers develop similar exhaust effecting devices, most notably Suzuki, Honda and Cagiva.

Again a lot of the engine and transmission components remained virtually identical, except for a beefing up exercise, which included sturdier shaft sizes within the gearbox to withstand the extra power.

The heritage of the 350YPVS's frame can be seen in the TZ racers of the late 1970s and early 1980s. The totally new, wide tubed chassis, employed a rising rate rear suspension system to control wheel movement and a generously proportioned box section swinging arm to resist rear end

flexing. Air assisted front forks, added a final touch to what was then the best Yamaha sports bike to reach mass production. Except for adding a full fairing for the 1985 season the same 350YPVS has remained in production ever since; it is today not made in Japan, but Brazil!

Finally in the LC story came the RD500LC; its 499 cc (56.4 × 50 mm) liquid-cooled 50 degree V-four motor made Honda and Suzuki rush back to the drawing board.

The RD500LC's engine featured not one but two crankshafts; at the same time Yamaha was clever enough to incorporate components first seen on the smaller 250/350LC models. Although its two cranks had a 50 mm stroke they were nevertheless copies of the unit fitted to the road-going twins.

The reed-valve induction differed from early Yamaha practice by having two distinct arrangements on the same engine. The front two cylinders were fed from reed blocks mounted directly onto the crankcases, and those at the rear were fed by the more conventional cylinder fitted components. This configuration allowed all four carburettors to be grouped inboard between the V of the cylinders, producing a very narrow, compact engine.

The crankshafts, both rotating in the same direction, drove a massive 14-plate clutch which in turn operated an engine speed double weighted balancer shaft. The gearbox, deviating from the normal splash fed variety seen on the LC twins, had a trochoid pump that delivered the oil to shafts, primary gears and sundry smaller items.

Equipped with a single electrically powered servo motor, all four YPVS

valves were mechanically linked. As well as a desirable reduction in production costs, the system also eliminated out-of-balance effects and ensured all valves operated in unison.

The magnificent RD500LC was a genuine road-going V4 two-stroke and not – as some pundits would have you believe – a 'Kenny Roberts Replica'. The powerplant was not adapted from a GP engine; many of its components came from the existing LC series, thereby maximising reliability. However, the cycle parts, unlike some engine bits, had been specifically designed for the V4. The frame was a wide, rectangular steel-tube affair with a single rising rate gas/oil suspension unit mounted horizontally below the engine. Everything smacked of performance; the bike was fitted with anti-dive front forks and triple ventilated disc brakes.

Although not fully related, the TZR250 carried the LC formula into the future – initially in parallel twin form and, as the 1990s dawned, as a V-twin incorporating much of the technology from similar factory racing machines.

The YPVS system is recognised as one of the classic innovations in two-stroke technology, if not in motorcycle engine design in general – a tribute to the technical reputation of the company that created it. Yamaha's rivals have had to work extremely hard to catch up.

Touring

After building a series of disappointing four-stroke parallel twins, beginning with the 650 XS1 of the late 1960s, Yamaha came up with the XS750 triple in autumn 1977.

The XS750 triple not only propelled Yamaha into the Superbike league, but it at last provided a serious touring potential. Featuring three cylinders, double overhead camshafts, 747 cc (68×68.6 mm) and shaft drive, the bike also had a modern set of cycle parts; (eg, a double cradle frame, triple disc brakes and cast alloy wheels). This well engineered package attracted excellent press and public reaction, and boosted sales.

Later E and F models had improved performance thanks to a higher compression ratio and an increase in engine revolutions from 7500 to 9000 rpm – maximum speed going up from 110 to 115 mph.

Even so, Yamaha chose to improve things still further, and for the 1980 model year overbored the three-cylinder engine by 3 mm to give 826 cc and create the XS850.

Developing 79 bhp at 8500 rpm and giving a top speed of 125 mph, the Yamaha triple was at last a match for rival four-cylinder machines of similar capacity and not a few 1100s. The extra 'umph' of the larger three-cylinder was not simply as a result of its extra bore size; the original

Left
Following the failure of the TX750 parallel twin, Yamaha introduced the XS750 with a 747 cc (68×68.6 mm) three- cylinder dohc engine. Capable of 110 mph the XS750 soon gained a loyal following as a long distance tourer, with a premium on comfort and smoothness. A couple of years after its launch the company overbored the engine by 3 mm to achieve 826 cc, and create the very similar, but more powerful XS850 for the 1980 season. Maximum speed increased to almost 125 mph

Next came a number of four-cylinder models including the XJ550/650/750 Seca series and the XS1100, before Yamaha came up with the excellent XJ900 in 1983. At its heart the newcomer had a four-cylinder dohc 853 cc (67 × 60.5 mm) engine, with shaft drive and five-speed gearbox. Early examples were slated by the press for poor high speed stability (traced to the fairing), but later models were much improved in this area and the model is still available in 1993. An unfaired 1985 model is shown

cam profile had been reworked for additional lift and dwell, the carburettors changed from Mikuni to Hitachi and the crankshaft beefed up. The gearbox was also given additional strength to cope with the increase in power, but the excellent Hi-Vo primary chain was retained.

A preload-adjustable front fork and rear shocks with 5-way settings provided handling and roadholding which was the equal of the lower powered 750 – a significant achievement, considering the extra performance potential.

The next stage of the Yamaha sports/touring story concerns the various short-lived four-cylinder models of the early 1980s: the XJ550, XJ650, XJ750 Seca and finally the ill-fated XJ650 Turbo. To be brutally honest none were particularly good, certainly when compared with what the company replaced them with – the FJ1100 and XJ900.

With both of these latter bikes, Yamaha created machines which have proved to be excellent in almost every respect, witnessed by the fact that they are still in production a decade later.

Launched in 1983, the original XJ900 had an 853 cc (67 × 60.5 mm) four-cylinder engine with dohc, five-speed gearbox and shaft final drive; it also came with a handlebar mounted fairing. This latter item was to prove the one bad thing about the bike, causing a frightening high speed weave, and was soon replaced (1984 model year) by an entirely new fairing mounted on the frame, a belly pan also being added at this time.

The XJ900 subsequently received an increase in engine size to 891 cc, this being achieved by larger 68.5 mm pistons, although maximum power

Above

For 1984 Yamaha designers came up with a brand new sports/tourer. Their brief was to produce a motorcycle that was capable of speeds in excess of 150 mph, could handle better than other touring-type machines, yet it would have to be just as happy swapping lanes in dense city traffic as it was eating up the miles on the open road. The result was the FJ1100, an across-the-frame four with sixteen valve head and 120 bhp plus power output. (FJ1200 shown.)

Left

The 1992 XJ900; by now engine capacity had increased to 891 cc (68.5 × 60.5 mm). Other details included a compression ratio of 9.6:1, 91 bhp at 9000 rpm and a maximum speed of 131 mph output hardly changed – from 89 to 91 bhp – rpm in both cases being 9000. But throughout its career the bike has retained 18 inch wheels, a dry weight of 218 kg (480 lb), air-assisted front forks with an equalizer tube linking each leg, twin De Carbon-type rear shocks, separate forged alloy handlebars, centrally mounted tacho and an oil cooler.

Air cooling and a two-valves-per-cylinder layout is simplicity itself, while routine maintenance is all but eliminated thanks to shaft drive, electronic ignition and automatic cam-chain tensioning. Together with its sensible purchase price and fuel efficiency, these qualities have kept the XJ900 selling steadily over the years.

But the real star in Yamaha's touring crown has to be the FJ1100/1200 series. Before the arrival of the first FJ1100 in 1984, Yamaha four-strokes had never really been taken seriously. With its two-strokes dominating much of motorcycle sport, coupled to the brilliant RD-range of two-stroke roadsters, public attention had been focused on these machines rather than the company's four-strokes. Moreover, the vast majority of Yamaha four-strokes were either in middleweight or lightweight capacities, or else were V-twins or shaft-drive tourers; good (or bad!), they were simply not glamorous enough to grab the attention of the motorcycle enthusiast and hold it to the detriment of the opposition.

The FJ1100/1200 models, with their mighty twin overhead cam, 20-valve four-cylinder engines and race-bred 'lateral' frames, changed all that for ever!

Performance Bikes summed up the FJ1100/1200 as, 'Not so much a sensible bike, more a bike built by a sensible designer'. In other words there was stacks of torque, lots of stability, plenty of comfort and a decent range between refills at the pumps. Only heavy steering and a lack of ground clearance marred this remarkable machine.

In creating the FJ1100, Yamaha saw fit to include several features from the Italian Bimota marque, notably 16 inch front and rear wheels, a massive 150 section rear tyre and, as for the overall similarity of the chassis, well . . . just how much Bimota was consulted is unknown.

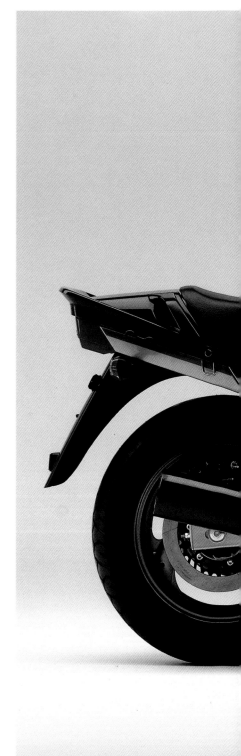

Two years on and the FJ1100 became the FJ1200; its air-cooled dohc engine now displacing 1188 cc (77 × 63.8 mm), maximum power increasing to a shade under 130 bhp. And, as with the XJ900, the FJ1200 has remained in production with very little change to this day. The only real update came in 1991 when the factory introduced the optional FJ1200A with ABS brakes and also several improvements to both versions including new orthogonal engine mounts with integral rubber dampers, increased rear shock travel, new control switches and thicker tank rails on the frame for increased rigidity

For its era the FJ1100 was one of the most stable bikes on the street; standard Yamaha 'Monocross' rising rate rear suspension, coupled with large diameter 41 mm front stanchions with anti-dive, gave a progressive, if slightly soft ride.

The straightforward, rugged, air-cooled motor was ideally suited to the sports/tourer role and cruising effortlessly on the highway. The FJ1100 could crack the standing quarter in a shade over 11 seconds, or plod along in top gear at a gentle 2000 rpm. Its top whack of over 150 mph was almost academic, because above all, the FJ1100 performed with an ease which others simply couldn't emulate. During 1984 the FJ1100 was Europe's biggest selling Superbike (although the Kawasaki GPZ900R changed all that the next year). Kawasaki's all-conquering 'Ninja', and the introduction of Yamaha's own FZ750 (see Chapter 11), forced the company to re-launch the big FJ concept with the FJ1200 for the 1986 season.

Engine displacement was increased from 1097 to 1188 cc, achieved by boring out the FJ1100 cylinders' from 74 to 77 mm while retaining the same 63.8 mm stroke. The result was an improvement right across the power range. More low-end power and mid-range torque plus the ability to pull a higher final drive ratio for increases in both cruising speed and maximum velocity.

Carburettor settings were altered to suit the new engine size and the air cleaner duct lengthened by 50 mm to reduce intake noise.

New stainless steel exhaust pipes were employed which reduced machine weight by 1.7 kg. The exhaust collector box of the four-into-one system was also revised and featured added internal partitions to reduce noise.

Actual engine modifications, except those necessary to achieve the extra cubes were detail ones aimed at further refinement and providing more reliability to the already proven FJ unit.

Most of the detail attention, however, was focussed on the transmission where widths were increased, the grove-pattern in the shifts drum modified and the collar pressed on to the fourth-gear pinion to strengthen it against the engine's extra torque.

Change to the frame and suspension was limited to providing the swinging arm pivots with needle roller bearings instead of bushes.

Aerodynamics had been improved by giving the FJ1200 a redesigned head fairing and engine undercowl. The fairing frontal area was reduced for superior air penetration but rider protection was also improved by integrating the front flasher units into 'hand guards' on the fairing sides. The fairing-mounted rear view mirrors had new, aerodynamic shapes and even their mounting stems were now of wind-cheating oval profile. The engine undercowl had been increased in size to cover the lower frame rails and more of the crankcase area, a major area of turbulence.

As well as the FJ1200 and 900, Yamaha also offered the FJ600, but this was always a bland bike and, consequently, a poor seller. The company sought to rectify this situation with the new-for-1992 XJ600S, more commonly known as the Diversion. The 598.8 cc (58.5 × 55.7 mm) dohc air-cooled engine employs a slant-block cylinder configuration inspired by the company's well-known Genesis concept. Yamaha claim this approach not only enhances weight distribution and handling performance, but also improves air management for more efficient engine cooling

Inside the head fairing were a new digital clock and an electric fuel reserve switch, both of which were aimed firmly at the touring brigade. And, choke operation was now via a pull-knob located in the fairing next to the reserve switch.

Finally there was a new, aircraft-type filler cap, and a restyled tail cowling made in a more scratch resistant material.

The next update came in 1988, with improvements aimed at increased protection and comfort (taller windscreen and reshaped dualseat and rubber mounted pillion footrests) new wheels including a 17 inch rim size at the front, revised brakes, and suspension changes (minor linkage differences at the back and the deletion of the anti-dive mechanism at the front) and new digital ignition and an electric fuel pump.

Except for graphics the FJ1200 remained unchanged until the 1991 model year, when it was significantly improved upon once more. Besides an ABS option (coded FJ1200A), vibration was cut to a minimum by new orthogonal engine mounts with integral rubber dampers – and for an even quieter top end the Yamaha design team fitted cylinder head fin spacers. Rear shock stroke was increased by 8 mm to 48 mm, giving a smoother rear suspension action; there were new control switches, while the perimeter frame now had thicker 50 mm tank rails for increased rigidity. There was also a second (taller) screen given to each new owner, offering even better protection.

Although Yamaha had offered the XJ600 for a number of years, it had never been very popular and, in an attempt to boost sales in this

important sector, the Division was introduced at the end of 1991. Although it was still powered by a 598 cc air cooled four- cylinder engine, there the similarity ended. The Division was very much a 2-valve air-cooled version of the Genesis-inspired slant block configuaration, not the old XJ series.

The Division soon proved popular, thanks in no small part to its value-for-money sports/tourer specification with excellent all-round engine and chassis performance combined with graceful, understated styling.

By mating proven technology with a functional yet attractive design Yamaha produced one of the best middleweight all- rounders ever built.

But if the Division was cheap and conventional, the next touring model from the company was just the reverse. The GTS1000, launched at the end of 1992, was not only an all-new machine, but one which represented a major leap forward in motorcycle design.

Centre of attraction was the radical Omega chassis with its single-sided swinging arm front suspension, its inherent anti-dive properties, which the factory claimed offered 'remarkable all-round handling performance even when carrying a passenger and luggage'.

The GTS was designed very much for the dual role of sports/tourer and much attention was paid to rider ergonomics, including seat/footrest/handlebar layout and the special fairing with its built-in air management system.

Designed to produce massive low/mid-range response for quick and safe overtaking and effortless touring, the 1002 cc (75.5 × 56 mm) 5-valve per cylinder motor features a host of technical gizmos. Not least the use of Electronic Fuel Injection (EFI), seen for the first time on a Yamaha, which ensures highly efficient combustion processes and instantaneous throttle response, while the 3-way catalytic converter reduces exhaust emmissions for a cleaner environment.

Fitted with ABS coupled to a large ventilated single disc with a new 6-piston 330 mm caliper at the front and single 282 mm disc at the rear the GTS seems to offer considerable technical progress... but is it what the motorcycling public wants? This is the big question, and only time will tell if Yamaha haven't over-reached themselves.

Yamaha's publicity machine billed it as a 'Ride in the Future'. They are of course referring to the new-for-1993 GTS1000 with its Omega chassis, hub link steering and electronic fuel injection. Only time will tell if the GTS really is the bike of tomorrow, or whether the buying public will give it the cold shoulder – the fate of many other futuristic concepts over previous decades

Vees

When Yamaha showed off its 1981 model line-up to American dealers in Las Vegas during the autumn of 1980 the big news was the company's first ever vee-configuration engined machines, the XV920 street bike sold in Europe with larger 981cc capacity as the TR1 and the XV750 Virago custom.

Projected star of the '81 Stateside range was the 920 cc, chain drive, monoshock, backbone-framed sports/tourer with which Yamaha hoped to steal some of the thunder from established V-twin manufacturers such as Ducati, Moto Guzzi and Harley-Davidson.

The other Yamaha V-twin, the 750 cc Virago, was based on the same 75 degree sohc V-twin engine and was styled in classic Stateside custom clothes with its stepped seat, pullback bars and forward-mounted pegs. It shared the same monoshock frame and many of the same engine components, but had a smaller bore size and also used shaft final drive, instead of chain as on the larger engined bike.

The Yamaha V-twins evolved in response to the firm's market research. This revealed that a large group of people existed who wanted a V-twin but wouldn't consider buying those V-twins already on the market.

From this market research, Yamaha Motor Corporation in the USA sent a request to Japan asking for a 750 cc V-twin that was to appear as the Virago. The XV920, however, stemmed from a request from European Yamaha importers who asked for their own V-twin but a machine more in tune with the sporting European riders. As a result, America got both vees.

A quick look at the history book shows that there was nothing new about V-twins, lots of companies from BSA to Zenith having tried the layout, including such famous marques as Husqvarna, Moto Guzzi and Vincent.

The first Yamaha ever to sport the vee configuration was the 981 cc (95 × 69.2 mm) TRI V-twin tourer of 1981 (there was also the stateside-only XV920 and XV750 Virago custom). The cylinders were angled at 75 degrees, which precluded the ultimate in smooth running achieved by Ducati's 90 degrees, but improved upon the narrower angle designs such as Harley-Davidson or HRD Vincent. An interesting feature of the TRI was its rear drive chain which was totally enclosed, running in lithium-based grease. This arrangement (if correctly maintained!) would provide for a greatly increased life for both the chain and sprockets. Some TRIs were even used for 'Battle of the Twins' racing as was this example at Daytona in the early 1980s

But Yamaha was different, because Yamaha wasn't building a V-twin for the same reason other companies built them. The earliest designs were natural because a V-twin could be developed from an existing single and it fitted perfectly into a bicycle-type frame. That's also why most of the V-twins in history have been from 45 to 50 degrees in cylinder spacing, though the Italians liked 90, 120 or even 72 degree spacing.

The Yamaha cylinders were angled at 75 degrees not because of the frame or because of engine balance, but because that provided the necessary amount of room for the carburettors in the middle and remained as compact as possible. The engine was constructed on a single throw crank, but with the con-rods running beside each other, not forked. Ball race bearings supported the crank at each end while the con-rods used plain bearing big-ends.

Cam chains were located at both ends of the crank, the offside (right) driving the cam for the forward cylinder and the nearside chain driving the cam on the rear cylinder.

Chain tension was provided by fully automatic cam chain tensioners. Each cam drove a single inlet and exhaust valve venting the modified hemispheric combustion chambers.

Another throwback to days past was the vertically split crankcase which also contained the five-speed transmission in the unitised cases.

Bore and stroke of the 920 was 92 × 69.2 mm, while the 740 had an 83 mm bore and the same stroke. The vastly oversquare engines redlined at 8000 rpm.

Because the forward cylinder acted as a frame member and was stressed, Yamaha saw fit to use a solid metal head gasket, and in order to eliminate oil seepage that would accompany the metal gasket the Japanese V-twin had external oil lines running from the banjo fittings on the crankcase up to the heads. The oil lines, vertically split cases and starter motor hung at the front of the engine in plain view and gave the assembly a slightly vintage appearance.

More modern bits included the 40 mm Hitachi CV carbs and electronic ignition that even employed electronic advance rather than the centrifugal advance used on most machines at that time or the centrifugal and vacuum advance Yamaha had fitted to their XS 1100 four.

Supplying the carbs with air was an unusual air intake system which drew air from under the seat, through a filter behind the engine and then passed the air through the massive backbone of the frame that led to intake ducts going to the carburettors. The air breather system was just one of several functions provided by the V-twins' frame.

The huge stamped and welded structure that was the frame's backbone might have brought back memories of the stamped-out chassis on Yamaha's ultra-lightweights of a decade earlier, but was, in fact, a highly efficient way to construct a strong frame. On the XV models the welded-up structure worked with the stressed engine, the monoshock and the air intake to solve design problems effectively.

Yamaha had a host of experience with monoshocks, going back to the original YZ250 works motocrosser of 1973. Single shock designs had become almost universal by the beginning of the 1980s in factory motocross machinery and even on works road racers. Single spring designs, certainly, were not new, but hadn't seen much use on the street bikes during the 1960s and 1970s, so Yamaha got credit for reinventing the idea.

In the pair of Japanese V-twins the monoshock was a combination coil and air spring with adjustable damping. The coil spring wasn't equipped with an easily adjusted preload, but the reinforced rubber air bag at one end of the suspension unit took care of suspension stiffness. The damping was adjustable with no less than 20 different positions. However, the remote adjustment knob only provided six adjustment positions. By resetting a remote cable different groups of six adjustments could be provided by the remote adjustement knob, but there was no easy way to have all 20 positions available without moving the cable!

To keep the damping adjustment constant at all temperatures, the damping adjustment was actually controlled by a tapered magnesium rod that screwed in or out as the damping knob was turned. Because the magnesium rod expanded and contracted with temperature changes, the damping hole became smaller at higher temperatures, retaining damping power even when the monoshock heated up with use.

Neither the 750 or 920 had a conventional final drive chain system. The smaller model used shaft drive, while the 920 had a fully enclosed drive chain running in a grease bath.

At each end of the chain a cast aluminium cover shrouded the sprocket. In between the shrouds were flexible heavy rubber tubes with bellow ends and inner rails moulded into the tubes to act as chain guides. The rubber, Yamaha claimed, was specially formulated to resist wear and work well with the heat and grease that lubricated the non O-ring 630 chain. Chain adjustment was made by tightening the chain adjusters to a given torque and backing off a prescribed number of turns.

Although the styling of the Virago was very much standard Stateside custom issue, the same could not be said of its street bike brother.

Although Yamaha didn't have much luck with pure street bike versions of their V-twins, the Virago custom series in 535, 750 and 1100 sizes was a different matter entirely. Over the years they have built a solid reputation and are today generally recognised as the best of the Japanese customs. The XV535 (shown) employs a 535 cc (76 × 59 mm) sohc air-cooled engine with two valves per cylinder. Running on a compression ratio of 9:1, maximum power is 47 bhp at 7500 rpm. Carburation is by a pair of Mikuni BDS 34 instruments and there is electronic ignition and a five-speed gearbox

On the XV920 the styling wasn't like anything else, with the strangest part of the machine being the rear end where the mudguard mounted onto the swinging arm and went up and down with the wheel, like a front 'guard. At the rear of the seat there was a small storage box, tail lamp and licence plate holder.

In Europe the XV920 was sold as the TRI with a larger capacity 981 cc (95 × 69.2 mm) engine, which produced 70 bhp at 6500 rpm and was capable of a shade over 110 mph. But although the Virago was reasonably successful, the XV920/TRI was soon axed due to low sales volume – victims of wierd styling and unexciting performance.

The next Yamaha V-twin was the hi-tech XZ550, a completely different machine to its predecessors with a brand new 70 degrees unit and a myriad of technical innovations – not all of which were to be successful in practice.

Heart of the XZ550 V-twin was its 552 cc (80 × 55 mm) engine with dohc 4-valve cylinder heads, 10.5:1 compression ratio and 0.65 bore/stroke ratio. Together with hi-tech gadgets such as liquid cooling, twin exhaust ports per head, pressure-fed wet sump lubrication, three-weight single-shaft engine balancer, twin Mikuni BD34 downdraught carbs, YICS (Yamaha Induction Control System), fully transistorised ignition with electronic advance and lightweight shaft final drive.

There were also a number of unusual technical innovations in the chassis. The most notable of these being the Trailing-Axle front forks. Other features included Monocross rear suspension, together with twin downtube, full cradle frame with its front downtubes running above the crankcase. This suspended engine mount design of frame was very unusual, but the Ducati Pantah V-twin of the same era used a similar type. A 4-1-2 exhaust system; adjustable, separate, handlebars and a distinct, if ugly, square styling exercise which extended throughout the whole machine completed the specification.

The XZ550 proved a costly error; not only was its performance unimpressive considering the complexity, but reliabilty in service proved abysmal, with major engine troubles occuring all too frequently. Once again, Yamaha was forced to withdraw a roadster V-twin after only a short sales life.

Ready to take off for the weekend is the XV535 Virago, Yamaha's entry-level custom. In this case, the sohc in-line v-twin develops 47 bhp at 7000 rpm allowing the bike to despatch the quartermile in a creditable 14.8 seconds; top speed is 98 mph. Streamline Motorcycles of East Dulwich, London, commissioned this one-off USAF-inspired paint job. Introduced for the 1988 season in the UK, the XV535 offers a choice of flat (shown) or pullback bars; shaft final drive, front disc and rear drum, dry weight 182 kg (401 lb)

In contrast, the Virago custom cruiser just kept going. In 1986 the engine capacity was upped to 981 cc (the same as the ill-fated TR1), later still it became the XV1100 with 1063 cc (95×75 mm). A notable change along the way being twin shock, instead of the single shock, rear suspension.

Meanwhile the company played a master stroke by introducing the 'compact custom', in the shape of a mini-Virago, the XV535. As with all custom models, the engine is the focal point of the machine and the designers of the XV535 had been extremely creative in achieving a 'big-motor' look from an inexpensive, middleweight (air-cooled) V-twin.

Yamaha did this by trimming down the lower half of the engine, using round crankcases and slimline side covers, and by accentuating the upper part of the motor with big squared-off cylinders and heads using fine-pitched cooling fins. These catch the eye and emphasise the V-twin look that is so vital to any Harley lookalike.

Technically, the 535 cc (76×59 mm) sohc 70 degree V-twin breathes through a pair of 34 mm Mikuni CV carbs, runs on a compression ratio of 9:1 and pokes out 47 bhp at 7500 rpm. Maximum torque is produced 1000 revs below this figure.

Eye-catching as well as efficient, the exhaust pipes curve and join on the offside of the machine and terminate in short, stacked mufflers. The whole engine unit is used as a stressed member of the backbone chassis to eliminate the need for downtubes, thereby accentuating the 'long and low' build of the XV535.

Like the other Virago custom models (which in 1993 not only comprise the 535, but 250, 750 and 1100), there is shaft final drive, together with a low (700 mm) seat height. (Not all models are offered in many countries).

Other 535 features include raked front forks with 150 mm of wheel travel and 36 mm stanchions; a single 298 mm floating caliper disc at the front and single drum rear brake; wire wheels (19 inch front, 15 inch rear); and a choice of short flat or pull-back custom hadlebars. Lastly, beneath the seat is the fuel tank, with fuel being pumped up to the carbs by an electric pump.

The 535 has proved a surprisingly popular bike, particularly on the British market, where in the past Japanese custom bikes have suffered an almost complete lack of interest; not so the middleweight Yamaha which has sold in solid numbers since its introduction to the UK for the 1988 season.

So far we have discussed only the various V-twins. Yamaha have also built V-fours in the shape of the XVZ12TD 'Gold Wing replica' highway cruiser and the incredible V-Max.

The mighty V-Max is certainly one of the most talked about motorcycles of recent years, being launched in America in 1985 and Europe a year

The 1992 Virago 1100 was a true custom iron, with plenty of 'street cred' having an engine powerful enough to handle long distance highway cruising with ease. The lusty 75 degree V-twin had a capacity of 1063 cc (95 × 75 mm) and a torque curve to provide low-down grunt equal to many race replicas. The 'gun-barrel' shaped silencers provided a traditional style, as did the drum rear brake and masses of chrome plate and polished alloy work

later (although British importers Mitsui chose not to stock it until 1992 – and then only in restricted 95 bhp form).

Its unashamed emphasis on horse power and forceful styling soon attracted the attention of both media and customer alike. What captured the imagaination was the news of test times on the American dragstrips previously unheard of for street bikes.

The very obvious focal point of the V-Max was the power unit. The engine was the same liquid cooled V-four that powered the XVZ12TD heavyweight luxury tourer. In the V-Max, however, it was propelling a machine of approximately two-thirds the weight and the logical results of this were obvious!

For its capacity the 1198 cc (76 × 66 mm) V-four is a compact engine. The two banks of cylinders are opposed at a 70 degree angle and the four constant velocity, direct downdraught carbs sit neatly between them. Cylinder heads have 4-valve combustion chambers and double overhead cams.

While most key engine specifications were the same as the XVZ12TD, power output had been increased by various upgrades in valve gear and carburation. The four Mikuni carburettors were of 35 mm bore – 1 mm larger than those on the motor's touring counterpart.

Inlet valves were increased in diameter from 29 to 30.5 mm and the exhaust valves from 24 to 25 mm. The stems of the valve were slimmed down to reduce reciprocating weight and improve the gas flow around them. Smoothness is an inherent factor of any V-four engine but Yamaha

engineers took this a stage further by the inclusion of a crankshaft counter balance shaft and rubber-mounting in the wide-cradle chassis. With the obvious emphasis on acceleration, much attention was paid to strengthening of the V-Max drive train in which power was transmitted to the dished rear wheel via a five-speed gearbox, hydraulic clutch and shaft drive.

Valve springs were stiffer and the 10.5:1 compression pistons and gudgeon pins were lightened so they could cope better with high-rpm running. Connecting rods and crankshaft were provided with improved hardening and heat-treating processes and both rod and crank bearings made tougher. Additional hardening also strengthened the various parts of the transmission.

The rear wheel has an enormous 150/90-15 tyre... dwarfing the 110/90-18 front assembly.

To retard the progress of this 258 kg (575 lb) giant, there are twin 282 mm discs at the front and a single disc of the same diameter at the rear. A round-section steel swinging arm and twin shocks were specified as Yamaha considered that these best suited the mean styling image that the V-Max projects.

Love it, or hate it, you have to accept that the V-Max is all set to become a classic in future years, if for no other reason than that it just can not be ignored!

Its mighty power delivery (at least in unrestricted form) is very deceptive; as much horsepower as many factory racing machines but delivered with such smoothness and in such quietness that riders new to the V-Max are strongly advised to acclimatise themselves to these characteristics before giving the bike a full turn of the right wrist!

The awe inspiring V-Max with its mighty 1198 cc (76 × 66 mm) V-four engine. That motor is the centre of attraction, featuring narrow-angle cylinders, liquid cooling, twin overhead camshafts and four valves per cylinder. Four downdraught carburettors are tucked away in the angle of the vee; there is electronic ignition, a crankshaft counter-balancer shaft to smooth out vibration and bright chrome-plated megaphone silencers to mould the unmistakable V-Max image. The fuel tank is hidden away in the mid-section of the bike, with a fuel pump supplying the carbs. Performance (at least in the unrestricted version) is mind blowing, but its forte is travelling at speeds up to 80mph – any faster and wind buffeting can be a real pain

Big Singles

Sometimes motorcycle manufacturers and the buying public have different ideas when it comes to the use of a particular machine. Take the XT500 (TT500 in the States) for example. Yamaha built the bike for casual on-off road riding; in otherwords a fun bike. Customers however thought they were getting a modern-day Gold Star, in other words a serious bike.

In addition the XT was a much better roadster than dirt iron.

The XT's problem lay mainly with its size, weight and frame geometry; an all-up figure of 138 kg (320 lb) was good news for a roadster but that mass off-road needed muscles that could only be learned from a Charles Atlas course.

It was big mainly because of the sohc 499 cc (87×84 mm) engine was tall, even though it was dry sump – with the oil carried in the frame. To achieve reasonable ground clearance and long travel suspension Yamaha engineers also made it taller than it need have been purely for street use. The fork legs protruded a full 39 mm($1^1/_2$ inches) above the top yoke, whilst the seat height of 940 mm (33 inches) was a stretch even for six-footers.

Unless you were really tall and well-built into the bargain fear and the XT went hand in hand if riding at anything above 15-20 mph off-road was to be attempted with any real verve (at least for a bog standard, showroom fresh XT).

As if accepting that weight was a major problem 1978 saw a number of improvements which collectively reduced this to 123 kg (308 lb). Much of this reduction came from manufacturing the 12 litre (2.3 imperial gallon) fuel tank from aluminium instead of sheet steel.

As for performance (on-road) Bike magazine achieved a top speed (sitting up) of 91.46 mph in a 1978 test.

But it was indicative of the mini-revolution that the XT500 started that not only did Yamaha go on to build a whole family of big four-stroke single cylinder trail and road bikes, but so did Honda, Kawasaki and Suzuki.

Yamaha followed the XT with the SR roadster, in 1978. The 87×84 mm overhead cam motor with its 5-speed gearbox and wet, multiplate clutch obviously had its origins in the XT, but had undergone a number of changes in its transfer from trial to pure street.

Most obvious of these was the substitution of polished aluminium engine side cases for the XTs lightweight matt black magnesium covers. The cylinder was bigger, with larger fins to improve cooling, while inside were reshaped ports with larger valves. The SR's Mikuni carb was more complex and 2 mm bigger than the XT's at 34 mm. Piston and rings had

The first of Yamaha's big four-stroke singles, the XT500 trail bike came to Europe in 1977 (followed shortly afterwards by the SR500 roadster). Centre-piece of the machine was its 499 cc (87 × 84 mm) sohc engine which pumped out 33 bhp at 6500 rpm. Its makers proclaimed it 'as a modern version of the classic big single'. Some observers likened it to the BSA's sadly missed Gold Star, while others said it didn't touch the long-stroke British unit for 'plonking' ability

been modified to cope better with long periods of high rpm running and the flywheels and crankshaft beefed up.

Vibration levels, compared to old British singles were reduced, but not entirely eliminated – yes both the XT and SR did vibrate.

The electrics were totally different from the XT's pretty minimal equipment. Instead of the dirt bike's flywheel magneto and six volt direct lighting, the SR had magnetically triggered electronic ignition and powerful 12 volt lighting. There was a valve lifter lever on the left bar and the kick indicator located on the offside end of the crankshaft which made it easier to get the piston in the right position – but both the XT and SR models could be a bitch to start with a hot engine (caused by fuel vaporisation).

With its 12 volt electrics, XT based motor disc/drum brakes and 162 kg (370 lb) dry weight the SR looked good on paper. Unfortunately it never

Left
On the street the successor to the SR500
was the SRX600 of 1986. This featured
not only the larger engine size of 608 cc
(96 × 84 mm), but a four-valve cylinder
head, square section frame tubing
monoshock rear suspension, triple disc
brakes and cast alloy wheel; plus of
course a more modern overall
appearance, perhaps best described as a
latter day café racer. Power increased
to 45 bhp, with maximum rpm
unchanged at 6500. Top speed was a
shade under 110 mph

received the rave reviews which greeted the XT; probably because with
only 33 bhp and 95 mph it wasn't very quick and the styling although
"clean" was boring . . . all in all, it added up to a bland bike, lacking in
any real character.

By 1983 Yamaha had expanded their big single range to include the
XT400 (399 cc – 87 × 67.2 mm), XT500 and XT550 (558 cc – 92 × 84
mm) and SR500. All these had motors based broadly around the original
sohc XT of the mid 1970s, but the XT550 was equipped with a 4-valve
cylinder head, twin carbs and YDIS (Yamaha Dual Intake System).

Then in 1985 came the 595 cc (95 × 84 mm) XT600 and XT600Z
Tenere models still with the familiar single overhead cam motor but
sharing the 4 valves and twin carbs of the XT550, together with
improvements to the suspension, braking and more modern styling.

For 1986 Yamaha saw fit to reintroduce the roadster theme (the SR500
had by now been dropped in most markets) with the SRX600.

Based on the familiar and proven XT600 motor, the SRX had a slightly
larger capacity 608 cc (96 × 84 mm) but retained the 4-valve head fed by
the twin carburettors of the unique YDIS. These two carbs operate like a
dual-throat automobile unit, varying the functional choke size with the
degree of throttle opening. The engine ran just on the primary unit at low
speeds, giving greater economy as well as instant engine response, thanks
to a direct-pull throttle operation. The secondary unit then came into play,
giving a progressively larger choke size until at full opening there is
approximately 25 per cent more intake area than with conventional

New for 1986, the XT350 used 4-valve technology and a state-of-the-art suspension to create a market leader for Yamaha in the 350 trail bike stakes. The 346 cc (86 × 59.6 mm) motor owed nothing at all to previous XT models in the under-500 cc bracket. A four-valve cylinder head, with narrow angle valves and double overhead camshafts, combined with Yamaha Dual Intake System (YDIS), provided the XT350 with a high revving unit and the power of many five hundreds

induction systems on just about any similarly-sized engines.

The secondary carburettor in the YDIS layout was a constant-vacuum unit that provided better high-rpm performance because it was vacuum-controlled by the engine pulses to give exactly the fuel supply that the motor demanded.

The SRX power unit was much more sophisticated than the earlier SR and had been designed to be smooth and easy in operation, together with low rider maintenance. Thus it featured a crankshaft counter-balance shaft, electronic ignition with preset timing and an automatic compression release linked to the kickstarter. This and the relatively low compression ratio of 8.5:1 meant that it was a much smoother, easier to live with engine than the earlier sohc 2-valve unit found on the SR.

Other details of its specification included twin-loop, full cradle chassis constructed in box-section, high tensile steel tubing and was based directly upon Yamaha's earlier road racing designs as used on the TZ range. The twin stainless steel exhaust pipes were routed out between the front downtubes and were tucked tight into the chassis to allow maximum cornering clearance.

Rear suspension was of the classic twin-shock variety, in keeping with the café racer image that Yamaha aimed to project. The tortionally-stiff, rectangualr section, steel swinging arm pivotted on needle roller bearings.

An important point for owners who carried out their own maintenance was that the two front chassis downtubes were detachable to facilitate engine removal.

The forks had 36 mm diameter stanchions, with 140 mm of stroke and were braced by an alloy bridge.

The twin 267 mm front disc brakes used opposed-piston calipers that generated higher braking forces than the conventional single-piston type. A single 245 mm rear disc was used and both wheels were cast alloy units.

Styling was much more attractive on the SRX than the old SR, with narrow, flat, clip-on bars and slightly rearset footrests which positioned the rider low over the flush-top 15 litre tank with its aircraft-type filler cap.

A centrally-mounted, white faced speedometer emphasised the classic look, with the small tachometer neatly offset in a separate bracket on the right.

Extensive use of aluminium parts gave the machine an air of lightweight and technical efficiency – something the SR500 could never achieve.

Many SRX600s found their way onto the race circuit as the basis for clubman's type single cylinder class racing machines.

Following the SRX600 came the XT350 (346 cc – 86 × 59.6 mm) and XTZ660 (660 cc – 100 × 84 mm) Tenere trail bikes. The former using the

Above
A 1990 XT600E (electric start); this dual purpose on-off road bike has a 4-valve sohc 595 cc (95 × 84 mm) motor with five-speed gearbox. The 21 inch front and 17 inch rear wheels are equipped with dual-purpose tyres and the long-travel (225 mm) front forks feature 41 mm stanchions, which together with Monocross rear suspension provide safe handling on a variety of surfaces. Braking is taken care of by a 267 mm drilled disc with Nissan twin-pot caliper up front and a 189 mm drilled disc and single pot caliper at the rear

Above

Fabulous Over-Yamaha single cylinder racer uses a tuned version of the XTZ660 Tenere engine. The liquid cooled 5-valve single features dual carbs. Bore and stroke measurement are 100 × 84 mm. The machine shown here is the one raced by the Japanese rider Shin-Ichro Ohura in the 1992 British HTH Sound of the Singles competition series. The other official Over teamster was Englishman Roger Bennett. With well over 60 bhp available the Over-Yamaha 660 single can reach speeds of around 140 mph; an amazing performance for an engine which started life in a trail bike

same engine technical features, but with a 6-speed gearbox, whilst the 660 has a 5-valve head, liquid cooling and YDIS (Yamaha Dual Intake System).

It could be said that the XTZ660 motor is one of the most technically efficient four-stroke single cylinder motors ever to reach mass production, and a credit to the Yamaha engineering team who conceived it.

Off-road sport

Motocross, trials, enduro – even sidecar cross, Yamaha-engined bikes have contested just about every facet of off-road motorcycle competition. However, motocross is, after road racing, the most important of all powered two-wheel sports.

Yamaha's first serious motocrosser didn't emerge until 1971, although their interest in dirt racing can be traced back to the early 1950s.

Four-time world champion Torsten Hallman was released by Husqvarna at the end of the 1970 season and Yamaha promptly stepped in and offered the Swedish veteran a contract not only to ride, but also to develop a complete range of new motocross machines.

Hallman did his job so well that, by the mid 1970s, the company was gaining GP success as well as producing a successful line of 'customer' motocrossers based on the works machinery.

Yamaha also recruited top class riders in the shape of Hakan Anderson, Heikki Mikkola and Bengt Aberg.

Another notable event was the appearance of the Monoshock chassis and a revolutionary suspension system, both invented by Belgian engineer Lucien Tilkin.

It was Hallman whom Yamaha had to thank for the three major moves that influenced its dirt bike programme; machine development, introducing Lucien Tilkin's ideas and – perhaps most important of all – judging to perfection the seventies boom in all forms of off-road sport that took place in the USA.

The first really proficient over-the-counter Yamaha customer motocrosser was the YZ125C of 1976 – much progress having been made from the original MX125 of three years earlier, which was little more than a race-kitted DT trail bike.

By 1980 the dealer network was able to offer a full range of YZ machines from 60 through to 465 cc.

Watercooling arrived for the 1981 model year in the shape of the YZ125H. Yamaha had been running works liquid-cooled 125s on and off for half a decade; now at last the fruits of this experience were available over the counter.

For 1982 Yamaha took the formula a stage further with the new YZ250J – a big brother version of the 125H. With liquid-cooling, YPVS and Monocross rear suspension it, like the 125, was at the leading edge of dirt racing technology. At the other end of the scale the 'men only' YZ465H grew into the even more muscular YZ490J.

The next major development came with the K series (1983) of YZs with

Above

British 500 cc motocross champion Neil Hudson storming to yet another victory in 1982 on his YZ490. At that time it was probably the most powerful production dirt bike racer in the world. The air cooled 487 cc (87 × 82 mm) single cylinder 2-stroke produced 60 bhp at 7000 rpm. There was reed valve induction, 38 mm VM Mikuni carb and a four-speed gearbox

Right

The 1982 500 cc British GP grid. Yamaha riders include Franco Picco (Italy), Jukku Sintonen (Finland) and Niel Hudson. Although the company didn't lift the title, Danny la Porte became the 250 cc world champion riding a YZ250

Above

A year later and the Swedish star Hakan Carlqvist is seen here in action during the 1983 500 cc British Motocross Grand Prix. Changes that year included a new link-type Monocross suspension system and a reduction in the total weight of the YZ490 achieved by careful attention to detail. The engine was now more compact; the cylinder head volume decreased and the fins on the cylinder and cylinder head reshaped. There was also a new front fork assembly and narrower handlebars

Left

Vivid action from the motocross races at Daytona in 1984. Rick Johnson (17) leads another Yamaha rider in the 125 cc race which he won. It is worth noting that while white and red are Yamaha's colours in Europe, across the Atlantic the livery is bright yellow

a revised version of the rising rate Monoshock which remained unaltered until 1986. Since then, a policy of continuous refinement, with upgraded versions appearing virtually every year, has maintained the competitiveness of the YZ series.

When it comes to versatility in the bike market, there's always something new under the rising sun. The Japanese are constantly expanding their horizons, manufacturing not only a vast range of street bikes, but a veritable army of off-road machinery. For Yamaha this means not only the YZ motocrossers but TY trials and IT enduro models. Yamaha has even dominated the little-known sport of sidecar cross, its XS series of parallel twin engines being widely used to gain a host of world championship titles.

Above

A 1985 production YZ80. This 79 cc (47 × 45.5 mm) liquid cooled flyer was every schoolboy's dream bike that year. Running on a compression ratio of 7.85:1, the tiny six-speed engine produced a staggering 22 bhp. Other details included a 26 mm carb and reed valve induction

Above right

A 1993 model YZ250. Technical highlights for the 249.8 cc liquid cooled reed valve machine include redesigned porting and combustion chambers for more engine torque, larger capacity radiator, upside-down front forks, new frame, redesigned linkage for the Monocross rear suspension, new Deltabox swinging arm, new chainguide, new seat, tank, side covers and rear mudguard for improved rider control

Right

Yamaha has also been a dominant force for many years in the specialised field of sidecar cross. This 1982 photograph shows the West German pairing of Josef Brockhausen and Herbert Rebece during the British Grand Prix at Leighton near Frome, Somerset. Their outfit is powered by an overbored XS twin cylinder motor in a British Wasp chassis

Above

Rob Sartin guns his IT250 through Forestry Commision land near Brandon, Suffolk during the Breckland Enduro; the first round of the 1993 British Championship series

Right

The legendary Mick Andrews competing in the 1984 Scottish Six Days Trial on his works supported 246 cc two-stroke single cylinder TY250. A decade before the same rider had gained the first ever Japanese win in the famous event – on a Yamaha of course

FZR – race replica

It may surprise many readers to discover that Yamaha began work on what was to emerge as the FZ/FZR project way back in 1977. What they were looking for was a four-stroke that could equal the power of two-strokes. Not unnaturally this entailed some fairly drastic thinking in the research and development division.

If you think 5-valve heads are a little over the top then consider that Yamaha's engineering team determined that seven valves was the optimum number for ensuring the highest volume of inlet mixture into a combustion chamber up to a rev ceiling of 20,000 rpm. The company even went to the trouble of building prototype single- and twin-cylinder engines to evaluate their ideas but ultimately came to the conclusion that there were problems with the technology of the period, particularly when it came to putting the idea onto the production line. Also, it transpired that seven valves in one combustion chamber were not best suited to the comparatively small displacement engines used in motorcycles.

Six valves were tried next but these suffered the major shortcoming of hot spots between the trio of exhaust valves. So how about five valves? Yamaha engineers soon discovered that the two exhaust valves were far enough apart to eliminate the hot spots and the three inlets still allowed a high intake volume.

As always with practical engineering, combining the right set of design features is the key to success.

The 5-valve layout meant that enough gas flowed into the (theoretically superior) lens-shaped combustion chamber to provide the potential for a relatively higher output than conventional 2 or 4 valve heads. Using a single plug – as opposed to the twin plug layout on the 7-valve prototype head – the biconvex chamber gave excellent flame propagation. In layman's terms, all the stuff shoved into the chamber is burnt quickly and efficiently. This had the added advantage of allowing high compression ratios with a slightly dish-topped piston. Yamaha engineers claim that this

Launched onto an unsuspecting world at the tale-end of 1984, the all-new FZ750 catapulted Yamaha into the age of the high performance, racer-replica four-stroke. The unique four-cylinder, 20 valve engine was chosen after extensive research, which included experimentation with 5, 6 and 7 valve layouts, and was fed by a bank of four vertically mounted, 32 mm carburettors. Revving to over 10,500 rpm, the 749 cc (68 × 51.6 mm) engine came with a host of features including a six-speed gearbox and hydraulically operated clutch. Top speed was 149 mph

set-up provides a 10 per cent power advantage over a 4-valve head using the traditional pentroof chamber.

In 1982 the Yamaha development team decided that they were on the right track and initiated a variety of projects all using the 5-valve head, lens-shaped combustion chamber and a capacity of 750 cc to comply with the upcoming regulations for F1 and endurance racing. Many varied configurations were tried along with such gizmos as hydraulic valve lifters. None of the separate project team leaders was aware of what his counterparts were up to, and when the time came to make a choice, the ever-popular across-the-frame four was given the go ahead.

Besides the 5-valve technology, the other radical feature of the final design was the 45 degree slant of the cylinders, which lowered the centre of gravity – a decision influenced by studying the latest trends in the configuration of automobile racing engines. The down-draught carburettors were borrowed from the ill-fated XZ550 V-twin.

The result was the FZ750, launched for the 1985 model year. At the time Yamaha stated that this bike would form the basis for its large capacity sportsters over the next decade.

On the FZ750 the carbs and airbox lived inside a dummy tank and the paths from the 100 mm bellmouths protruding into the airbox were routed past the flat, plastic sides down the inlet tracts to the valves in straight and equal lengths.

Similarly, the exhaust ports and header pipes were straighter than would have been possible with an upright engine or, as Yamaha was keen to point out, with a V4. Another advantage of this arrangement was that the airbox was in still, cool air.

The balance of the liquid-cooled dohc 749 cc (68 × 51.6 mm) engine displayed Yamaha's old obsession with narrowness. The generator was located on top of the gearbox behind the cylinders and chain driven off the crankshaft. Not content with that, the electronic ignition trigger was on the nearside crank web and the sensors mounted on the crankcase halves. As a result the FZ's crankcases were 9 mm narrower than the XJ400 air-cooled four and as slim as a V4. The width across the barrels had been kept to a minimum by a hybrid wet/dry liner system which directly watercooled the top section of the liners only.

Steve Parrish on his Team Loctite FZ750 in the 1985 Isle of Man TT Production race. The 45 degree forward inclined cylinder bank permitted the advantages of superior centre of gravity, centralised mass and equal front and rear weight distribution, straight intake and exhaust paths, together with the very latest in four-stroke technology. Unfortunately all this was not quite enough to defeat the all-conquering Suzuki GSXRs that year

Left

The FZR750R (OWO1) was introduced for the 1989 model year and employed a new version of the famous FZ/FZR four cylinder, 4-stroke motor, with its inclined cylinders and liquid cooling. The new unit had very over-square dimensions with 72 mm bore and 46 mm stroke and was derived from the 1988 YZF750 factory endurance racing engine. The redesign had been so thorough that it was virtually a new engine when compared to the previous year's production FZR750 motor. There was also a comprehensive racing kit

Below left

With its 5-valve cylinder head, slant block and efficient inlet and exhaust systems, the FZR1000 quickly established itself as one of the world's finest motorcycles. For 1989 it was improved further still. A brief summary of the updates for that year include a higher redline thanks to a lighter valve train, more displacement (1003 cc instead of 989 cc), a higher compression ratio and redesigned combustion chamber with straighter inlet ports, bigger carburettors, a reduction in frictional losses with thinner piston rings, and the addition of the EXUP exhaust control system

The only truly conservative engineering feature concerned the centrally located camchain, which makes things awkward when a replacement is required. To justify this Yamaha's engineers stated that the torque loadings caused by a camchain on one side of the power unit would be unacceptable when the engine was race kitted.

The valve train design saw a cam lobe for every valve – direct operation without rockers – and hollow chrome molybdenum camshafts. Valve lash adjustment consisted of shims under the bucket seats. Yamaha claimed that after the running-in period the valves are virtually maintenance free, helped by the entire circumference of the cam lobes being carburised, not just the lifting surfaces. The bucket-and-shim arrangement meant that load-bearing areas were bigger than with conventional tappets; also the bucket tops were manufactured from sintered material for improved wear resistance. On the hot side the valve seats were also made from sintered metal (as one would expect in an air-cooled engine running at much higher temperatures than the liquid-cooled FZ), thereby minimising heat-induced wear.

The valves themselves were each equipped with a single, oil-tempered, silicon-chrome spring, and had to cope with less severe lift than on a 4-valver due to superior gas flow.

This all added up to a very tough valve train. After all, automobile designers have managed to do away with the sort of regular adjusting that the bike industry had grown accustomed to, so there was no reason why Yamaha shouldn't follow this route.

The balance of the engine looked quite conventional when compared to the top end. There was a six-speed gearbox and a hydraulic clutch but otherwise the FZ mostly followed standard 'UJM' (Universal Japanese Motorcycle) practise, though there were some nice little touches such as the positive oil feed to the gearbox, pressure equalising holes inside the crankcases to cut down on pumping losses, and baffles in the sump to reduce the lubricant's tendency to create crankshaft drag.

In contrast to the engine the FZ's chassis looked almost ordinary. It was a box section affair, but made of steel not alloy and derived from the YZR500 grand prix racer of the early Kenny Roberts era. The front downtubes were detachable for easier engine removal and the top tubes splayed dramatically for better access to the top end.

The box section swinging arm was aluminium alloy, pivoting on needle roller bearings and controlled by the rising rate Mono Cross system. Up front the 39 mm fork stanchions were manufactured from tubing with a robust 3 mm wall thickness.

A trio of opposed piston calipers with semi-metallic pads took care of the braking. The six-spoke cast alloy wheels were 16 inch up front and 18 inch rear.

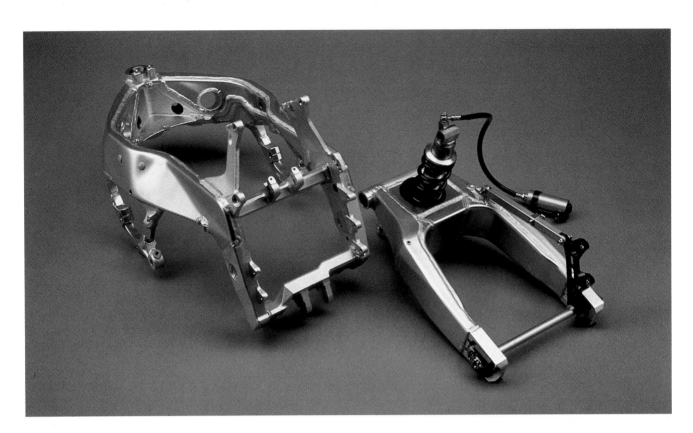

For 1989 there was also a new version of the aluminium Deltabox frame developed from the factory's YZR racers. That year saw the deletion of the dual front downtubes from 1988. The engine now bolted directly to the frame at the cylinder head, at the top of the upper case and, like before, at the rear. By making the engine a stressed member overall frame rigidity and strength was significantly improved. Also new was an increase in the swinging arm pivot diameter from 16 to 20 mm

Another important step occurred in 1989 with the FZR600, replacing the air-cooled FZ600. The newcomer followed the 'Genesis Concept' of inclined engine and Deltabox frame that had made the larger FZRs such an outstanding success. The FZR600 also gave the company a machine with which it could at last take on the best in the competitive Supersport 600 racing class. Interestingly, the 599 cc (59 × 54.8 mm) engine employed 4 instead of 5 valves per cylinder

As for the fuel tank, this occupied the rear of the dummy tank cover and had straight-sided walls dropping down to just above the gearbox. The change in the C of G from empty tank to full tank was therefore a mere 20 mm straight down as opposed to a much larger drop and a rearwards shift as with a conventional layout.

As for power output, this broke the 100 bhp barrier (106 bhp at 10,500 rpm to be exact); but the most impressive feature was the torque. The almost flat torque curve saw 8 kg/m (63 ft/lb) at 8000 rpm, truly excellent figures indeed.

The basis was there for a long-running series of super-sports bikes and it made sense for Yamaha to develop the theme further.

The first step came two years later with the FZR1000 Genesis.

The focal point of most new models is the engine, but not so the FZR. This time it was the chassis – namely the Deltabox frame. The Yamaha press kit at the world launch described 'Delta' as a Greek letter meaning 'triangular in shape', and when the FZR was viewed from above this statement became obvious.

It was public knowledge that Yamaha had been playing around with alloy frames on their 500 cc GP bikes since the middle of 1979, when Kenny Roberts complained specifically that he was far from happy with the then current steel job.

Originally these were little more than box-section alloy replicas of their ferrous predecessors; in fact the earliest examples were painted black to fox the opposition. The prototypes proved extremely prone to cracking. This was remedied by employing more gussetting and progressively thicker-walled tubing, until by mid-1981 the frame was actually heavier than the steel one it replaced!

The 1982 Yamaha V4 YZR500 saw the first beam-like frame, the head stock being connected to the swinging arm pivot by an almost straight member of vast torsional stiffness. By the following year it was recognisably 'Deltabox' and subsequent refinement saw it tidied up as confidence grew in its inherent rigidity.

The 1987 FZR1000, together with the TZR250 two-stroke twin of the same era, were the first two production Yamaha models to make use of this race-developed frame technology.

How did it work in practice? Well, compared to the steel framed FZ750, the FZR1000 had a seat height which immediately encouraged fast riding. The clip-ons were also lower (and considerably narrower!) than the 750 and placed under – rather than over – the yokes, and the foot pegs were set further back. With the 750 the rider felt perched on top and divorced from the road, whereas the 1000 placed you in the bike right at the heart of the action.

The important 'numbers' were similar for both models: the 1000 had a

Above

A 1990 FZR1000 EXUP, the top one-litre sports bike that year, with razor sharp handling and dynamic performance. This is the second generation FZR1000, between the original non-EXUP model, and the single headlamp version which appeared in 1991

Left

Team Mitsui-Loctite rider Rob McElnea riding a 750 OWO1 at Donington Park, World Superbike series, April 1990

mere quarter-inch more trail, fifth of a degree less rake, 0.8 inch less wheelbase and one inch more front wheel than the 750. In theory then, this should have meant similar steering characteristics, but the 1000 felt both more willing into a turn and more neutral through it. The difference was really down to the 'friendliness' of the larger model's riding position. When it came to high-speed stability and steering precision, the FZR was the clear winner.

As for the rest of the chassis, the Genesis was surprisingly standard. The front forks were short, braced and substantial, with 41 mm stanchions and devoid of unnecessary extras. The rising-rate rear end was tidy and equally spartan, with only a threaded pre-load collar to fiddle with.

There were also massive, hollow-spoke wheels and high twin 320 mm discs at the front and a single 267 mm rear disc. As for the engine, this was virtually pure FZR750, but bored and stroked from 68×51.6 mm to

Above

Chain drive and Michelin low profile rubber put the power down; a single disc brake helps wash the speed off. The YZF750R has hollow wheel axles and spokes, and race-bred swinging arm rear suspension with Öhlins shock

Left

An uncompromised, though street legal version of its endurance racing brother, the 1993 YZF750R featured the latest refinement of Yamaha's Deltabox frame and computer controlled EXUP technology

75×56 mm, giving 989 cc. During development the company encountered vibration problems that were only cured by burning a lot of midnight oil in designing pistons and con-rods that were lighter than their smaller predecessors.

Despite the increased bore size, the FZR's block was no wider than the 750's. This was made possible by using parts of 'siamesed bores' – best described as a sort of figure '8' with two cast iron liners side by side. Carburettor sizes were up by 3 mm to 37 mm, inlet valves 2.5 mm to 23.5 and exhaust valves 2 mm to 25 mm. Combustion was improved by revising the chamber shape and decreasing the valve angle, actual figures being 22 degrees to 30 degrees depending upon which valve you were looking at.

Taking a leaf out of Suzuki's book, the FZR1000 employed under-piston oil-jet cooling besides its FZ-derived liquid cooling; the size of the radiator

Left
Phil Read Junior competing in the
Bantam Racing Club's 600 Supersport
race at Snetterton, 13 March 1993. The
FZR600 is capable of 95 bhp and 145
mph in standard trim

and oil cooler were also increased. The clutch, final drive and crankshaft
had also been given more strength and the five gears of the Genesis'
transmission were inherently sturdier than the six of the FZ750.

Certain markets (including the USA) got the 'full power' 135 bhp
version, but many European countries (including Britain) received the
restricted 125 bhp model – either way top speed was breathtaking at 161
and 165 mph respectively.

The next Yamaha four-stroke development was EXUP (Exhaust
Ultimate Power Valve).

This was a special variable exhaust valve fitted to the exhaust system's
collector box, and operated by a computer-controlled servo motor which
performed according to engine rpm.

The valve effectively regulated exhaust pipe-end conditions, thereby
controlling the exhaust gas pressure wave produced during inlet and
exhaust valve overlap.

EXUP has the advantages of improving intake efficiency, reducing fuel
consumption and allowed engineers to improve low-mid range torque
without having to sacrifice peak rpm performance.

The first Yamaha production model to feature EXUP came in the spring
of 1987, and was the 'Japan only' FZR400R. This motorcycle had a 399 cc
(56 × 40.5 mm) Genesis-design, 16 valve, liquid cooled, across-the-frame,
four-cylinder engine featuring narrow-angle valves, straight inlet tracts
and compact combustion chambers for optimum power output, while the
block was inclined forward at 35 degrees.

The FZR400R evolved not only for the street but also the race track, featuring many race-qualified components and EXUP as standard. There is a healthy market for racing bikes in Japan; in 1986 no fewer than 23,450 (that's right, over twenty-three thousand) new Japanese licensed racers lined up on the grid – a 35 per cent increase on 1985. The majority of these competitors started in the novice TT Formula and SP-Formula classes – both open to production based machines – so it made good business sense to develop the FZR400R.

There were three developments for the 1989 model year: an EXUP-equipped FZR1000, the FZR750R (OWO1) and the new FZR600.

Not only did the one-litre bike get the advantages of the EXUP system, but it also received a new Deltabox frame. Gone was the dual front downtubes of the previous model's frame. The engine now bolted directly to the frame at the cylinder head, at the top of the upper crankcase and, as before, at the rear. By making the engine a stressed member, overall frame rigidity and strength were greatly increased.

This more compact design made it possible for a 10 mm shorter wheelbase and a new 26-degree fork angle improved responsiveness to turning inputs for more accurate steering control. The front fork stanchions were increased from 41 to 43 mm, while at the rear there was a modification to the linkage arms, increasing shock absorber stroke from 50 to 70 mm.

There were also changes to the rear wheel diameter (from 4.50×18 inch to 5.50×17 inch); new 4-pot opposed-piston calipers at the front with the top piston larger than the bottom one (33.96 and 30.23 mm respectively); and new larger diameter, hollow wheel axles and swinging arm pivot. The front axle diameter increased from 15 to 17 mm, the rear from 17 to 20 mm, and swinging arm pivot from 16 to 20 mm – these features coming directly from the YZR racers. Other more minor changes were made to the streamlining, FAI (Fresh Air Intake) system and electrical equipment.

Finally to the engine, which had a higher redline thanks to a lighter valve train, more displacement (1002 cc – 75.5×56 mm), a higher compression ratio (12:1), thinner piston rings (to reduce frictional losses) and a redesigned combustion chamber with straighter inlet ports; the carbs were also bigger (38 mm), and of course the motor now had EXUP.

In 1987 and 1988, Yamaha won the Suzuka Eight Hours race, one of the toughest events in the world for four-stroke racing machines. Now riders around the world were to have the opportunity to buy a replica of the victorious YZF750 factory racer. The new FZR750R (OWO1) was a limited production 'replica', fully street legal and produced in enough numbers for homologation in both FIM Superbike and Formula One world championships.

It employed an ultra-short stroke 72×46 mm, 749 cc, Genesis-inspired

A 1993 FZR600, which except for its colour scheme, remained unchanged from the previous year. Maximum speed is 146 mph with roadholding and braking to match. It was very much a case of keeping faith with a proven winner, both for street and track use

engine. This shorter stroke effectively reduced piston speed so the engine could rev higher without the limiting factor of piston failure. The corresponding increase in bore meant a larger-diameter combustion chamber, with space for larger valves.

Besides the bigger valves there were very light, immensely strong titanium-alloy connecting rods, plus lightweight pistons with only two rings (one compression and one control).

Other details included flat slide carbs, digitally controlled electronic ignition, a 15 mm lower cylinder block (due to the shorter stroke), which in turn allowed an alteration in the engine inclination from 45 to 40 degrees and a 20 mm shorter wheelbase for quicker steering.

Racing demanded a close-ratio six-speed gearbox, special wet clutch and the removal of the radiator cooling fan. In addition, the bike was upgraded with alterations to the Deltabox frame, an Öhlins racing rear shock, 17 inch hollow-spoke wheels and naturally enough production racer-type Michelin radial tyres of wider and lower profile.

There was also an extensive (and expensive!) factory race kit. This included forged pistons, an exhaust to be used with the existing EXUP system and inlet camshaft and carburettor-setting parts to adjust the air/fuel mixture required for the increased power output. The racing emphasis was underlined by the fitment of a new aluminium fuel tank of 20 litres capacity and by accommodation for the rider only — there was no pillion seat. No 'racer replica', the OWO1 was designed from the outset for racing, rather than street, use.

The third newcomer for 1989 was the FZR600, which replaced the ageing air-cooled FZ600 in the increasingly important middleweight division. The FZR600 followed the 'Genesis concept' of inclined engine and Deltabox frame which had made the bigger FZRs such a success. But it didn't boast the 5-valve cylinder heads, sharing the 4-valves per cylinder arrangement of the smaller FZR400 instead.

Displacing 599 cc from a 59×54.8 mm bore and stroke, this twin cam engine delivered a shade over 90 bhp at 10,500 rpm. The motor was kept on the boil with a six-speed 'box in which the fifth and sixth gears used reverse-taped dogs (back-cut) for positive engagement and improved reliability.

Much of the technology was identical to the other FZRs but the components came in smaller sizes: 38 mm fork stanchions, dual 298 mm front discs with a single 245 mm rear disc together with 110/70 V17 (front) and 130/70 V18 (rear) tyres.

Since its introduction, the FZR600 has been a leading contender for the keenly contested Super Sport 600 production-based racing category, even though Honda, Kawasaki and Suzuki have all attempted with varying degrees of success to muscle in on the scene during the FZR600's production life.

In the bigger capacity classes there have been a number of changes. For a start the FZR1000 received upside-down front forks and new slanted single headlight fairing (both 1991), while the OWO1 was replaced by the exciting YZF750R (street) and YZF750SP (Superbike racing) models for 1993.

Both these later machines offer new levels of engine and chassis performance that have made them the bikes to beat in the extremely competitive 750 category class, giving road and competition riders the chance to experience the power of a true factory racing machine.

Together with the FZR400, 600 and 1000 models, these two newcomers look like re-affirming Yamaha's position as the world's leading manufacturer of pure sporting motorcycles with inherent 'raceability', rather than the sports/touring image displayed by the vast majority of similar bikes from rival Japanese manufacturers.

Four-stroke parallel twins

'Someday you'll own a Yamaha' extolled the company's American advertising campaign in the early 1970s; but for many purchasers of Yamaha's four-stroke parallel twin models built in that era, it was very often a case of wishing that they hadn't taken this advice to heart.

But to Yamaha's credit they made amends in later years by playing harder, coming from behind and going for the four-stroke market that they had neglected for so long.

The very first of the series was the XS1 of 1969, which Yamaha saw as a way of cashing in on its similarity to the British-made vertical twins that had gained such a loyal post-war following, particularly in North America, where Triumph enjoyed world-record status from their speed exploits over the Bonneville Salt Flats of Utah.

As the British factories disappeared, by the beginning of 1969 only Triumph and, to a lesser extent, BSA and Norton survived in any real strength. Yamaha imagined that its 653 cc (75 × 74 mm) overhead cam XS1 would simply take over as a readily available twin exhibiting the old British values of power and straightforward design engineering, plus one or two of its own – notably, reliability and sophistication.

What Yamaha didn't realise was that the British bikes had character, even if less durable components such as bulbs, bolts and nuts would work loose or shatter with often clockwork regularity.

The Triumph's, Norton's and BSA's also had brand loyalty garnered over decades and handed down from father to son, and if all this wasn't enough the XS1 had major handling flaws; its frame, suspension and tyres were simply awful – something which couldn't be said of British bikes. The result was a motorcycle which was not well received, even though Yamaha revamped, renamed (XS650) and constantly improved it, until finally, by 1980 it was almost perfect ... but by then obsolete.

Next came another less than successful design, the much heralded TX750.

If the XS650 took a long time to mature, the TX750 didn't even get that chance; being axed within a few short months of its 1972 launch.

It was designed as a larger, smoother, more powerful stablemate to the XS650. Larger it was, at 763 cc (80 × 74 mm) and certainly smoother with its vibes tamed by a complex balancer system. Named 'Omni-Phase' by the company, the system consisted of two counter rotating weights housed at the rear of the crankcase and acting to cancel out the amplitude of the crankshaft. Unfortunately in achieving this task they also sapped much of the engine's power! This meant that the TX was slower than the smaller

Yamaha's first 4-stroke twin was the XS1 followed by the XS650. Both were distinguished by a 653 cc (75 × 74 mm) overhead cam engine and a five-speed gearbox; equivalent British vertical twins still had pushrods and a four-speed 'box. The original 1972 XS650 (shown here) came in a metallic gold finish and had drum brakes both front and rear. The design (albeit in custom form) was to survive until the early 1980s and during its life carried out a wide variety of tasks, not only on the street, but including American flat track racing and even sidecar motocross

Left
Honda were first to offer a 400 class four-stroke vertical twin; then came Kawasaki, followed by Yamaha, just ahead of Suzuki. Launched in 1977, the XS400 was a bored and stroked version of the earlier, short-lived 360. Its 392 cc (69 × 52.4 mm) engine provided adequate, if not exciting performance

Below left
Launched in late 1990, the TDM850 features a torquey 849 cc (89.5 × 69.5 mm) 10-valve parallel twin engine housed in a Deltabox chassis. With its comfortable riding position, unique styling and excellent motor, the TDM falls between a race-replica and trail bike. Yamaha class it as 'New Sport'. This is the 1993 model distinguished by its redesigned exhaust system with twin chromed silencers

twin, struggling to achieve 100 mph, which wasn't part of the plan.

Next came another expensive failure, the XS500. This was launched in Europe during 1975 and its double overhead cam 498 cc (73 × 59.6 mm) motor was capable of 48 bhp at 8500 rpm, which on the street meant 106 mph.

Inevitably compared with Honda's 500T, the XS was the more sophisticated design, with not only dohc, but 4-valves per cylinder and a balancer shaft (of different design to the ill-fated TX) operating off the crankshaft.

On paper it looked a fine bike, but acceleration was only moderate, not helped by its 209 kg (460 lb) dry weight. The much older, cruder and cheaper Triumph Tiger 100 had equal, if not better performance.

The first Yamaha four-stroke twins to sell in any numbers were the XS250/360/400 family which first appeared in early 1976 (360).

All these featured roughly the same single overhead cam motor with 180 degree crankshaft and six-speed gearbox. Displacement details are as follows: XS250 (248 cc – 55 × 52.4 mm); XS360 (360 cc – 66 × 52.5 mm); and XS400 (391 cc – 69 × 52.4 mm). It was simply a case of boring out the cylinders to achieve each increase in capacity.

The last of the series, the XS400 was totally revised for the 1983 model year with the engine capacity changed to 399 cc, by increasing the stroke by 1 mm. In addition, it was given a dohc top end, totally revised bottom end, new frame (with no front downtubes), single shock rear suspension, new eight-spoke cast alloy wheels, new upswept exhaust system and completely revised styling with square headlamp and instruments.

Other features included YICS (Yamaha Induction Control System) and, by mounting the alternator up behind the crankshaft, the engine was considerably narrower than the earlier models.

But once again Yamaha misjudged the market and this expensive update was wasted thanks to poor sales and the subsequent axing of the model.

There then followed a period where this engine configuration was not produced by Yamaha, that was until the emergence of the XTZ750 Super Tenere, and the new TDM850 which appeared in 1991, made use of the slant-angle cylinders, liquid cooling and the 5-valve per cylinder technology first seen in the four-cylinder FZ750 back in 1985.

What was different was the fact that these were both twins – the XTZ (749 cc - 87 × 63 mm) and the TDM (849 cc – 89.5 × 67.5 mm). If nothing else, the XTZ and TDM models prove that Yamaha does not give up easily – if you don't succeed at the first attempt, try, try and try again.

Specifications

Model	YSD1	YDS2	YDS3	YDS5
Year	1959–62	1962–64	1964–67	1967–69
No. cylinders	2	2	2	2
Bore (mm)	56	56	56	56
Stroke (mm)	50	50	50	50
Capacity (cc)	246	246	246	246
Compression ratio (to 1)	8.0	7.5	7.5	7.3
Power (bhp)	20	25	28	29.5
@ rpm	7500	7500	8000	8000
Valve type	ts	ts	ts	ts
No. gears	5	5	5	5
Front suspension	teles	teles	teles	teles
Rear suspension	twin shock	twin shock	twin shock	twin shock
Tyre front size	18	18	18	18
Tyre size rear	18	18	18	18
Dry weight (kg)	138	155	158	148

Model	DS6	DS7	YM1/2	YR1/2
Year	1969–70	1970–73	1965–69	1967–69
No. cylinders	2	2	2	2
Bore (mm)	56	54	60	61
Stroke (mm)	50	54	54	59.6
Capacity (cc)	246	247	305	348
Compression ratio (to 1)	7.3	7.0	7.5	7.5/6.9
Power (bhp)	30	30	26/31	36
@ rpm	7500	7500	7500/7000	7000/7500
Valve type	ts	ts	ts	ts
No. gears	5	5	5	5
Front suspension	teles	teles	teles	teles
Rear suspension	twin shock	twin shock	twin shock	twin shock
Tyre front size	18	18	18	18
Tyre size rear	18	18	18	18
Dry weight (kg)	138	138	155 (YM1) 142 (YM2)	152 (YR1) 157 (YR2)

Model	R3/R5	RD 250	RD 350	RD 400
Year	1969–73	1973–79	1973–75	1976–79
No. cylinders	2	2	2	2
Bore (mm)	61/64	54	64	64
Stroke (mm)	59.6/54	54	54	62
Capacity (cc)	348/347	247	347	398
Compression ratio (to 1)	7.5/6.9	6.7	6.6	6.2
Power (bhp)	36	30	39	40
@ rpm	7000	7500	7500	7000
Valve type	ts	ts	ts	ts
No. gears	5	6	6	6
Front suspension	teles	teles	teles	teles
Rear suspension	twin shock	twin shock	twin shock	twin shock
Tyre front size	18	18	18	18
Tyre size rear	18	18	18	18
Dry weight (kg)	148 (R3) 134 (R5)	146	148	151

Model	RD 250 LC	RD 350 LC	RD 500 L	TZR 250
Year	1980–83	1980–83	1985–87	1987–90
No. cylinders	2	2	4	2
Bore (mm)	54	64	56.4	56.4
Stroke (mm)	54	54	50	50
Capacity (cc)	247	347	499	249
Compression ratio (to 1)	6.2	6.2	6.6	5.9
Power (bhp)	35.5	47	87	50
@ rpm	8500	8500	9500	10,000
Valve type	ts	ts	ts	ts
No. gears	6	6	6	6
Front suspension	teles	teles	teles	teles
Rear suspension	monoshock	monoshock	monoshock	monoshock
Tyre front size	18	18	16	17
Tyre size rear	18	18	18	17
Dry weight (kg)	133	137	180	128

(Details apply to original parallel twin with conventional cylinders)

Model	DT1	DT400	XT500	XS1/XS650
Year	1968	1975–78	1976–81	1969–82
No. cylinders	1	1	1	2
Bore (mm)	70	85	87	75
Stroke (mm)	64	70	84	74
Capacity (cc)	246	397	499	653
Compression ratio (to 1)	6.8	6.4	9	8.4
Power (bhp)	21	27	27.6 (1976)	33 (1981) 50
@ rpm	6000	5000	5500 (1976)	6000 (1981) 7000
Valve type	ts	ts	sohc	dohc
No. gears	5	5	5	5
Front suspension	teles	teles	teles	teles
Rear suspension	twin shock	twin shock	twin shock	twin shock
Tyre front size	19	21	21	19
Tyre size rear	18	18	18	18
Dry weight (kg)	105	124	138 (1976) 123 (1981)	205

Model	TX 750	XS 500	XS 360	XS 750
Year	1972	1975–77	1976–77	1977–80
No. cylinders	2	2	2	2
Bore (mm)	80	73	66	68
Stroke (mm)	74	59.6	52.4	68.6

apacity (cc)	743	498	358	747
ompression ratio (to 1)	8.0	8.5	8.7	8.5
wer (bhp)	63	48	50	64
rpm	6500	8500	*	7500
alve type	sohc	dohc	sohc	dohc
o. gears	5	5	6	5
ront suspension	teles	teles	teles	teles
ear suspension	twin shock	twin shock	twin shock	twin shock
yre front size	19	19	18	19
yre size rear	18	18	18	18
ry weight (kg)	224	201	177	232

Model	XS 400	XJ 650 Turbo	XS 1100	XJ 900
ear	1978–83	1983–84	1978–81	1983–
o. cylinders	2	4	4	4
ore (mm)	69	63	71.5	67 (1983) 68.5 (1986+)
roke (mm)	52.4 (1978) 53.4 (1983)	52.4	68.6	60.5
apacity (cc)	392 (1978) 399 (1983)	653	1101	853 (1983) 891 (1986+)
ompression ratio (to 1)	9.0	8.2	9.0	9.6
wer (bhp)	32 (1978) 45.5 (1983)	90	95	97 (1983) 92 (1986+)
rpm	7500 (1978) 9500 (1983)	9000	8000	9000
alve type	sohc (1985) dohc (1983)	dohc	dohc	dohc
o. gears	6	5	5	5
ront suspension	teles	teles	teles	teles
ear suspension	twin (1978) monoshock (1983)	monoshock	twin shock	twin shock
yre front size	18	19	19	18
yre size rear	18	18	17	18
ry weight (kg)	177 (1978) 166 (1983)	230	261	218 (1983) 219 (1986+)

Model	XJ 1100	XJ 1200	Diversion	GTS 1000
ear	1984	1985–	1992–	1993–
o. cylinders	4	4	4	4
ore (mm)	74	74 (1985) 77 (1986+)	58.5	75.5
roke (mm)	63.8	63.8	55.7	56
apacity (cc)	1097	1097 (1985) 1188 (1986+)	598	1002
ompression ratio (to 1)	9.5	9.7	10.0	10.8

Power (bhp)	125	125 (1985) 130 (1986+)	61	100
@ rpm	9000	9000	8500	9000
Valve type	dohc	dohc	dohc	dohc
No. gears	5	5	6	5
Front suspension	teles	teles	teles	swing axle
Rear suspension	monoshock	monoshock	monoshock	monoshock
Tyre front size	16	16	17	17
Tyre size rear	16	16	18	17
Dry weight (kg)	242	227; 232 (with ABS)	187	246; 251 (with ABS)

Model	XT 350	XTZ 750 Sup. Tenere	XTZ 660 Tenere	FZ 600
Year	1986-	1989-	1990-	1986-89
No. cylinders	1	2	1	4
Bore (mm)	86	87	100	58.5
Stroke (mm)	59.6	63	84	55.7
Capacity (cc)	546	749	660	598
Compression ratio (to 1)	9.0	9.5	9.2	10.0
Power (bhp)	45	70	48	69
@ rpm	6500	7500	6250	9500
Valve type	dohc	dohc	sohc	dohc
No. gears	5	5	5	6
Front suspension	teles	teles	teles	teles
Rear suspension	monoshock	monoshock	monoshock	monoshock
Tyre front size	21	21	21	16
Tyre size rear	17	17	17	18
Dry weight (kg)	155	195	169	186

Model	FZ 750	FZR 1000	FZR 600	FZR 400
Year	1985–87	1987–	1989–	1987–
No. cylinders	4	4	4	4
Bore (mm)	68	75.5	59	56
Stroke (mm)	51.6	56	54.8	40.5
Capacity (cc)	749	1002	599	399
Compression ratio (to 1)	11.2	12.0	12.0	12.2
Power (bhp)	107	125*	90	66
@ rpm	10,500	10,000	10,500	12,500
Valve type	dohc	dohc	dohc	dohc
No. gears	6	5	6	6
Front suspension	teles	teles	teles	teles
Rear suspension	monoshock	monoshock	monoshock	monoshock
Tyre front size	16	17	17	17
Tyre size rear	18	17	18	17
Dry weight (kg)	205	209	179 (1992 model)	160 (1992 model)

* In restricted form

→

Model	TR1*	XZ 550	XV 535	V-Max
Year	1980–82	1982–83	1989–	1986–
No. cylinders	2	2	2	4
Bore (mm)	95	80	76	76
Stroke (mm)	69.2	55	59	66
Capacity (cc)	981	552	535	1198
Compression ratio (to 1)	8.5	10.5	9.0	10.5
Power (bhp)	70	64	46	130**
@ rpm	6500	9500	7500	8000
Valve type	sohc	dohc	sohc	dohc
No. gears	5	5	5	5
Front suspension	teles	teles	teles	teles
Rear suspension	monoshock	monoshock	twin shock	twin shock
Tyre front size	19	18	19	18
Tyre size rear	18	18	15	15
Dry weight (kg)	207	189	182	262

* Sold in USA as XV 920 (920cc – 85mm × 69.2mm)   ** Restricted on certain markets to 100 bhp

One of the brand new 1993 YZF750s being put through its paces; absence of streamlining gives a rare sight of its impressive technical details